This volume offers both a wide-ranging overview and detailed analyses of high, intermediate, and labor-intensive technologies that are appropriate for developing countries. The first section (Issues) looks at appropriate technology in the context of history, political development, and economics, with a focus on the relationship between developed and less developed countries. In the second section (Cases) the contributors explore the effects of technology transfer and dependence on such key nations as the Philippines, South Korea, and the People's Republic of China. Two concluding studies examine weapons production in less developed countries and the potential influence of direct satellite broadcasting on technology transfer in the Third World.

Appropriate Technology: Choice and Development is a valuable source of materials for both the beginning student and the specialist.

Related titles in Duke Press Policy Studies

A Society for International Development
Prospectus 1984
Edited by Ann Mattis

Modeling Growing Economies in Equilibrium and Disequilibrium
Edited by Allen C. Kelley, Warren C. Sanderson, *and* Jeffrey G. Williamson

The World Tin Market
Political Pricing and Economic Competition
William L. Baldwin

Also of Interest

The Growth of Economic Thought (revised and expanded edition)
Henry William Spiegel

Appropriate Technology: Choice and Development

Edited by Mathew J. Betz, Pat McGowan, *and* Rolf T. Wigand

Duke Press Policy Studies
Durham, North Carolina   1984

© 1984 Duke University Press, all rights reserved

Printed in the United States of America on acid-free paper

Library of Congress Cataloging in Publication Data
Main entry under title:

Appropriate Technology

  (Duke Press policy studies)
  Includes bibliographical references and index.
  1. Appropriate technology—Addresses, essays,
lectures. I. Betz, Mathew J. II. McGowan, Pat,
1939–   . III. Wigand, Rolf T. IV. Series.
T185.A7 1984   338.91   84-8187
ISBN 0-8223-0573-9

To Dr. Paige E. Mulhollan, Executive Vice President, Arizona State University, whose vision, encouragement, and support made possible our seminar and this book.

Contents

Introduction   ix

I. Issues

1. What Technology Is Appropriate?   *Mathew J. Betz*   3

2. Intermediate Technology and Development   *Warren Adams*   15

3. The Political Economy of Intermediate and Appropriate Technology   *Pat McGowan*   31

4. The Transfer of Technology to Third World Countries: Political Problems and International Ramifications   *Werner J. Feld*   49

II. Cases

5. From Dependency to Self-Reliance: An Evaluation of China's Experience of Technology Transfer   *S. Ivy Lang*   67

6. Nuclear Power in the Philippines: Technological Choice and Dependence   *Robert L. Youngblood and Melba D. Solidum*   83

7. Intermediate Technology in Newly Industrialized Countries: Two Cases from South Korea   *Martin H. Sours*   99

8. Weapons Production in Less Developed Countries: A Possibility for Integrating Technology with Development?   *Daniel J. Bohlin*   109

9. Direct Satellite Broadcasting: A Case for Appropriate Technology?   *Rolf T. Wigand*   121

Notes   145

Index   157

Contributors   162

Tables

5.1. Sino-Soviet trade, 1950–79   71
5.2. China's import dependence compared with that of selected countries and the world, 1975   77
9.1. Current and planned Third World regional and domestic satellite systems   130
9.2. Satellites operated and owned by international organizations or by developed countries capable of reaching remote areas   132
9.3. User cost comparison for educational television   139

Introduction

This book is a collection of selected papers from an interdisciplinary graduate seminar at Arizona State University that addressed the question of technology in economic development, placing particular emphasis on intermediate and appropriate technology. As such, it is a suitable resource for an upper division or graduate course concerned with the opportunities and problems involved in technology transfer.

From the beginning it was our desire to investigate issues in the broadest social, political, and cultural contexts and not to dwell solely on technical matters. It was not our wish to produce a series of case studies or how-to-do-it approaches to intermediate technology. Publications of this nature are readily available from many organizations, some of which are represented in this volume. Rather, it has been our goal to prepare selected materials that would be of interest not only to the technical practitioner but also to social scientists and to those making decisions on investment, capitalization, project development, and implementation in Third World countries.

The papers were initially presented to the seminar in the spring of 1981, but all were revised in 1982 and 1983. The current situation in the Third World further underlines the need to address the issues examined by our authors. Economic conditions continue to weaken; and although there may be some progress in stemming population growth, there seems to be little success in moderating the flow of people from rural to urban areas.

The financial condition of many Third World countries is now critical in the extreme. Debt service payments to both private and public lenders are being met with the utmost difficulty, often requiring rescheduling to avoid the most serious economic consequences. The fact that much of this financial strain is the result of the use of high capital technology in the development schemes of the sixties and seventies should be obvious to all.

More labor-intensive technologies would appear, on the surface at least, to provide substantial advantages in meeting development goals. This proposition has been recognized by many authors, including those represented in this volume. The importance of the capitalization rate in technology transfer has been carefully and clearly identified as to its foreign exchange implications, impact on employment rates, acceleration of urbanization, and role in debt service costs. It is our hope that the following papers will go beyond this traditional view, not because it is unimportant, but to better understand why appropriate or intermediate technologies either have been ignored or, in many cases, have not worked in one application after being successful in another.

In his contribution Mathew Betz emphasizes the fact that the technological base of labor-intensive technologies is thin, and that in most cases labor-intensive technologies by definition require management and entrepreneurial skills to achieve technological changes and to motivate the large numbers of people concerned. Warren Adams then presents a concise history of the experience gained by the Intermediate Technology Group created by E. F. Schumacher. Although substantial knowledge gaps still exist about intermediate technology, the Group more recently has focused on institutional problems related to the socio-political system that are needed to support, maintain, and replicate existing experiences. Pat McGowan directly addresses aspects of the international economy that work against a more favorable environment for the acceptability of labor-intensive technologies. These include the role of the international state-system, and he provides an interesting history of the role of "dominant powers" in technological invention and innovation since the United Provinces in the 1600s. The application of these ideas to contemporary technology problems identifies the need to focus greater efforts in the area of "soft technology." Werner Feld clearly articulates many of the conceptual and practical political and legal problems of international technology transfer. He uses these as the basis for exploring the status and future of the United Nations Conference on Trade and Development.

The second section of the book offers a wider geographic focus. S. Ivy Lang gives us a concise review of the changing attitudes toward technology transfer in modern China, concluding that China's record is not unique and that there is much in China's experience to be learned by other nations that are currently striving to solve similar development problems. Then, Robert Youngblood and Melba Solidum describe the process by which nuclear energy was selected as an important future power source in the Philippines. The role of international dependence is investigated as a primary force in this dubious technological decision. Martin Sours presents two interesting case studies of manufacturing technology transfer in South Korea. This chapter provides counterpoint to much of the material on intermediate technology that most often centers on technologies employed in rural development. It also gives an illuminating account of the role of Japan as the technology-source country. Daniel Bohlin presents one of the least considered forms of technology transfer—that of weapons production. Although the role of the military as a labor source is sometimes addressed in economic development literature, few investigations have been concerned with the role of military weapons, their production, use, and maintenance in the development of technological skills. Rolf Wigand concludes the volume with a most exciting and stimulating account of the opportunities for considering a highly sophisticated technology as appropriate to Third World development. This is the technology of direct satellite broadcasting in which the marginal cost of distance becomes virtually zero. The implication of this technology for countries without sophisticated ground-based communication infrastructures is an appropriate capstone for the volume.

The book's consensus is basically one of guarded optimism which recognizes that technology transfer will inevitably continue. Whether more labor-intensive technologies will be accepted and successfully implemented will be governed by a myriad of factors, including the traditional economic and technical ones. The successful implementation and growth of intermediate and appropriate technologies require much greater attention to the sociopolitical context in which the technological experiments are implanted.

The editors hope that readers will find these chapters both illuminating and thought-provoking, and that they will be an aid to those who wish to pursue the problems related to economic development of the Third World. The global development problem is immense in both human and physical terms. We attempt through this book to illuminate one small but vital area of the problem.

Tempe
July 4, 1983

Mathew J. Betz
Pat McGowan
Rolf T. Wigand

I. Issues

1. What Technology Is Appropriate?

Mathew J. Betz

During the past half decade the term appropriate technology has become used widely throughout the United States, and there is a growing body of publications addressing numerous aspects of the subject. The term itself can be interpreted broadly. Appropriate technology can be applied not only to the developing countries, but to Western Europe and the United States as well. Philosophically, appropriate technology implies that the decision makers should have the wisdom to devise, plan, evaluate, and select from a range of technical solutions to a given problem. Furthermore, it suggests that the decision makers' selection should be based on a broader range of criteria than has been true in the past. Appropriate technology emphasizes the need to include social as well as economic criteria in these decisions. It advocates a greater number of economic indicators than are addressed in the normal economic feasibility study.

Stated in another way, appropriate technology can be defined as providing technical solutions that are appropriate to the economic structure of those influenced: to their ability to finance the activity, to their ability to operate and maintain the facility, to the environmental conditions involved, and to the management capabilities of the population. There are numerous criteria to be examined. Appropriate technology challenges all. Not only the engineer, technologist, and economist, but the sociologist, anthropologist, historian, and others need to become involved in the evaluation and selection procedures of technological decision making.

In an extensive publication prepared for the National Academy of Science, Richard S. Eckaus develops the following criteria for appropriate technology: (1) To maximize product output, (2) to maximize the availability of consumer goods, (3) to maximize the rate of economic growth, (4) to reduce unemployment, (5) to encourage regional development, (6) to reduce balance of payments deficits, (7) to provide greater equity in income distribution, (8) to promote political development, and (9) to improve the quality of life.[1]

I would add at least the following to the above, although conceding that the list is still far from comprehensive: (10) To reduce the population flow to urban centers, (11) to provide an adequate national food base, (12) to be as consistent as possible with the indigenous social structure, and (13) to build upon and preserve the indigenous cultural continuity and heritage.

It is obvious that appropriate technology criteria are themselves in conflict. This is the real world situation where no solution, technical or otherwise, will improve all factors affected by a project. One strength of appropriate technology is that it can identify negative aspects of a situation as well as those that can be improved. This strength should lead to more rational decision making since the positive and negative aspects can be compared as tradeoffs. Such decisions presume, however, that the criteria and their relative importance can be agreed upon. Paradoxically, the advantage or strength of appropriate technology also may be identified as its weakness. It may fail at times because it cannot be all things to all people. Considerable delays, which may in the end be disadvantageous to all, may be encountered in the extensive analysis and evaluation required where so many diverse criteria are involved. Decision making becomes highly complex because of the number of criteria involved and can lead to disagreement as to which components have priority over others.

Definition of Appropriate Technologies

Because of its broad definitional base, one can honestly say that in the appropriate conditions any technology from virtually toolless hand labor to earth satellites can be appropriate. As applied to developing countries, the range is not substantially decreased. Not only must labor-intensive technology be developed and encouraged, but satellite and microwave technology is bringing communication and global integration to the entire world with physical, political, and social implications that are positive as well as negative.

Broadly categorized, technology as applied to developing countries can be classified in terms of capital intensiveness (capital investment per employee) and in terms of whether the technology is indigenous to the country or transferred from a foreign source. For the most part, technologies that have been transferred from other countries have been moderately or highly capital intensive. These are either technology implantations or technology substitutions.

Technology implantation involves the direct transfer of the most advanced technologies with little or no modification. The use of orbiting satellites, microwave transmissions, radio, and television are only a few of such transfers that have had and will continue to have substantial impact and are certainly justified. Similar transfers involve modern industrial equipment and/or plants.

Technology substitution would include many capital-intensive technologies, most of which have originated in the developed countries, that substitute for existing technologies in the receiving country by updating their equipment or procedures. These technologies tend, however, not to be in the forefront of technological development. In most cases they are established technologies that emphasize productivity, capital investment, quality of product, and similar criteria. The provision of a modern sugar refinery to replace a thirty- to fifty-year-old refinery and the substitution of modern textile mills for more ancient varieties are pertinent examples. Characteristically, then, these technologies update the status quo, usually with greater capital investment per employee and a possible decrease in the number of employees. The quality control of the product is normally improved, which may make the product more salable in the world market than it had been as described in the chapter by Sours on Korea in this volume.

The category of labor-intensive technologies would include all types of technology, whether indigenous or transferred, where the capital investment per employee is substantially less than that employed in the developed countries. Although this category does appropriately contain traditional, highly labor-intensive technologies, it also would contain the range of technologies defined as intermediate technology by Schumacher and others. Many of these technologies are being indigenously developed and refined in a number of countries. Additional intermediate technologies are being perfected within developed countries and then transferred. The possibility with the most promise is the transfer of such technologies from one Third World country to another.

In these cases the basic economic concern is the substitution of local labor costs for foreign exchange equipment costs. This is an important factor for the oil-importing developing countries that are extremely deficient in foreign exchange. An additional obvious advantage of such technology is the decrease in the number of underemployed in the area. The underemployment problem in the world need not be documented. New or improved intermediate technologies have implications in terms of slowing the migration from rural to urban areas and, perhaps most important of all, in making countries as self-sufficient as possible in basic food commodities. It is naive to believe that intermediate technology can completely stem current rural-to-urban migrations. Such technologies are unlikely to be that effective or pervasive throughout any country. The more labor-intensive intermediate technologies are but one part of the range of alternatives available. Nevertheless, I hope to demonstrate that greater weight should be given to some of the decision criteria that would tend to encourage and justify experimentation and implementation of such technologies.

The idea of more labor-intensive technologies, especially as applied to rural areas, goes back to the colonial era. Village industries were encouraged in India before the 1930s.[2] Gandhi's writings and philosophy were intimately tied to labor-intensive technologies. More recently the writings of E. F. Schumacher have given emphasis to some of the advantages of labor intensity. Schumacher's thoughts have had significant impact on British and American readers and have led to the development of intermediate technology groups in Great Britain and other countries.[3]

Needless to say, not all agree with Schumacher. Eckaus tends to classify the advantages of Schumacher's proposals under his (Eckaus's) criteria of improvement of quality of life and specifically limits them to rural village development. Eckaus states that the advantages of such intermediate technologies are limited and seems to imply that they fly in the face of major trends throughout the world by resting almost entirely on a village-oriented life-style.[4]

Many authors have indicated the barriers to the acceptance and implementation of the more labor-intensive technologies. Most of the technology available to the Third World is the technology of the developed world. This technology, generally based on an economy with high rates of employment and high wages, has survived and prospered by looking at a narrow range of solely economic criteria that emphasize minimal cost of production, capital investment, and quality control. It is not surprising that such an environment encourages capital-intensive investment projects. Project feasibility studies in the Third World, conducted by American or European consultants, tend to have a capital-intensive bias, intentionally or unintentionally. This bias may be unintentional in the sense that these consultants are experienced and technically more comfortable with capital-intensive alternatives. They may be biased intentionally because of the realization that the developed countries, for the most apart, are the source of the capital items at the time of implementation. More labor-intensive technology usually involves fewer expatriates in its implementation and operation.[5]

As much of the development in most developing countries is either carried out or channeled through the government (at least most of that which receives attention), the selection of projects is highly influenced by the local bureaucracy. It is clear that for a given amount of capital to invest, the larger the capital component of each project, the fewer the total number of projects. Thus, from the bureaucratic standpoint, it may be advantageous to have a limited number of projects that can be realistically followed and controlled rather than a very large number of low capital-investment projects, which may be difficult to control by a central agency.

Another built-in bias for capital-intensive projects may be the tariff and tax structures of the country. These structures can lead to what is sometimes called subsidized capital. Such policies tend to undervalue capital investment in relation to other alternatives. In some cases capital goods are imported free of tariff, while duty is charged on consumer or intermediate goods. In other cases capital may be provided through the central government at low rates, thus subsidizing those who wish to place their resources in capital investment as opposed to investment in labor components.

An additional bias for a limited number of high capital investment projects is the shortage of management talent in many Third World countries. This may not really be a bias as much as a fundamental fact.[6] Thus, many feasibility studies, both in the public and private sector, tend to select capital-intensive alternatives.

The desire of the private entrepreneur to capitalize is seldom addressed in the literature, other than in terms of lowering unit production costs to maximize profits. Increased capital investment also may increase capacity and quality control, which may broaden the potential market. If this is supported by governmental capital subsidization, so much the better from the entrepreneur's standpoint. Additionally, one of the factors that is virtually never mentioned and that may be a consideration even at a very low level of capitalization is the fundamental truth that the capital investment belongs to the capitalist. If an individual has given resources, he generally can select a technology that will place some of these into capital investment in plant and equipment, some into operational costs, and some into labor costs. There may be the feeling that operational and labor costs are dispersed, whereas investment in plant and equipment is under the control and ownership of the investor. In other words, expenditure in capital tends to be a long-term asset, whereas expenditures in operations and labor are not. The importance of this factor as a psychological bias is surely underrated. Its impact becomes even greater when dealing in economies with high inflation rates.

The preceding are basically economic and structural limitations to the implementation of labor-intensive technology. There are others: (1) There is little or no economic incentive for private investment in an invention of low capital investment tools. (2) There are no well-defined research and development laboratories, either in the private or public sectors, in most developing countries. (3) There is an absence of the entrepreneurial system represented by the individual firm in the private economy. (4) There tends to be a lack of marketing and distribution systems for those labor-intensive technologies that are known. (5) There is a weak financial infrastructure to support innovation systems for labor-intensive technologies because of the lack of private incentives and public interest. (6) As previously stated, there also is a lack of management expertise in the implementation of such techniques.

Thus, there might seem to be little justification for pursuing and placing greater emphasis on the more labor-intensive technologies. Although the large-scale agricultural development project continues to be broadly desired in many countries, most are currently developed with highly capital-intensive techniques. The Rahad project in the Sudan presents one such example. Some 90,000 people will be settled into this agricultural scheme. Thus far, it appears that it is likely to become a highly successful project providing a substantial number of Sudanese citizens opportunities to become part of their modernizing society. It is estimated that each person will achieve an income of approximately $200 per year ($1,200 per six-person family). Nevertheless, when the 15,000 families inhabit the 300,000 acres, it appears that the project investment will amount to $25,000 per family.[7] It would seem unreasonable that capital-intensive projects such as this one are the only solution for increasing agricultural output. The most important justification of a labor-intensive emphasis should be its positive impact on an increased number of individuals rather than its total impact on the national economy's gross output.

Development of Labor-Intensive Technology

The literature on labor-intensive technologies is a rapidly expanding one. The research of such technologies can be developed from one of four primary sources. First, there is the revival of older technologies. These technologies were used to build the original manufacturing plants and basic infrastructure of the developed countries and were much more labor-intensive than technologies today. The construction of railroads in the American West is a classic example. There are those who feel that a reintroduction of such technologies to Third

World countries is appropriate. Much of the appropriate technology literature, both as directed toward economic development or toward a different life-style, is fundamentally based on this idea. It has the advantage of being easily and quickly identifiable and having demonstrated success in the past. Thus, it can produce many good ideas at a very low cost. It is often the approach of the "instant expert," a breed not unknown in this field. It has the disadvantage of a lack of depth once the initial inventory has been conducted. It also has a very strong psychological disadvantage for the receiving country that is being advised to use century-old techniques. Further, this approach has the technical disadvantages that the tools are no longer available and that maintenance and repair parts that may have been readily available when the technology was broadly used fifty years ago are not available today. One would have to establish a fifty- to one hundred-year-old production and technology base to broadly implement such technology. Where this can be done on a very local level with minimal manufacturing and very simple, maintainable tools, it can be effective. If the technology identified has much sophistication, however, these technical limitations are difficult to overcome. This type of solution must be one of the first investigated in any comprehensive research effort, but it should not be overemphasized or become the primary intellectual base.

The second source of labor-intensive technologies is the adaptation of current technologies to a smaller scale for implementation in the receiving country. This is done daily as new plants and techniques are introduced. Even substantial manufacturing plants often are not the scale that would be built in the developed world. The adoption of smaller, lightweight tractors as opposed to heavier, commercial ones is an example of such technologies.

A third very important source is the adaptation and improvement of indigenous technologies. The fact that the populations concerned have existed within their geographical location and physical environment for centuries, if not millennia, is often overlooked. The methods and technologies developed certainly have been successful in those conditions. Thus, the "folklore" methods of doing things may form a fundamental base of knowledge from which to develop improvements and modifications. Furthermore, any new technique or tool must be introduced into the social and cultural environment that exists. This is the same environment that has successfully adopted or adapted traditional techniques. Why they work and how they work is fundamental in either improving existing techniques or in the insertion of new techniques into the cultural and social milieu. It would seem obvious that the less disruptive a new technique to its social and cultural environment, the more likely it is to be readily adopted. Labor-intensive technologies often apply to large numbers of people in open environments rather than to limited numbers in the closed environment of a factory.

The fourth source of more labor-intensive technologies, which currently is least used, is simply the invention of new technologies that are labor intensive rather than capital intensive. The difficulties in achieving this goal are substantial. We have built educational and research facilities aimed at the invention and development of tools and techniques that tend to be more capital intensive rather than less so. The difficulties to be addressed in looking at the opposite side of this coin cannot be overemphasized. Those who say, in effect, supply enough money and my engineers and scientists will develop labor-intensive tools are probably, naive. It is more than a question of direction and financial resources. It is a question of the whole flux of the education system from preschool through postgraduate levels. All of these factors, in both the developed and the developing countries, tend to build in biases, capabilities, and value structures that make it very difficult to invent and conceptualize low capital investment alternatives. This is not to infer that it cannot or should not be done. If labor-intensive techniques are to be broadly or even moderately effective, such research emphasis *must* be developed. It is probable, however, that such technology will be developed by an integration of institutions in the United States, Europe, and the Third World, for most of these techniques require extensive development and implementation projects that are best done in Third World locations.

The use of secondary education as a part of the development process may be more important than university education. The development of skills, many in labor-intensive techniques at the secondary level, represent a partial solution to one of the persistent problems in many countries—that of secondary school training. The village polytechnics developed in Kenya, although not without their problems and detractors, have been an important experiment, and their graduates usually have had a higher employment rate than other secondary school graduates.[8]

Rural Development

Much of the research and literature thus far are concerned with labor-intensive technologies applied to an individual actively related to rural development. The majority of these data focus on agricultural production, and many fine experiments, papers, and books have been developed. These technologies (many of them older technologies) involve small-scale projects, often with only a single person providing the source of power. The concept applies where separate projects can be implemented independently with effective results. The development of various methods of pumping water for agriculture is a common example. The mechanization of drying cassava and the development of small-scale distillation of palm wine in Nigeria are further examples.[9]

The development of a wide range of agricultural tools has been undertaken,[10] and the improvement and development of building materials, especially for residences and storage facilities, have been much researched. For example, hand-operated cement block manufacturing has been highly successful in a number of countries. Francis Stewart gives an excellent description of the factors and complex interrelationships encountered in cement block making in Kenya.[11] He also mentions the interesting phenomena that high-income housing tends to use both capital-intensive building materials, such as concrete panels and blocks, and high-quality, labor-intensive materials, such as wood or stonework. The labor-intensive manufactured (intermediate technology) materials tend to be used by the low-income homebuilder.

Stewart also has an excellent presentation on the technology of maize grinding in Kenya, comparing hand grinding to hammer mill grinding and roller grinding. Emphasizing the scale of labor-intensive technology, Stewart states that the capital investment per employee for roller milling of maize is two hundred times that for hand milling. Even when taking the different output into consideration, the capital investment per ton of output is still ten times greater for roller mills than for hand mills. Nonetheless, there seems to be a growing demand for roller mills in Kenya. This demand may be due to the fact that roller mills produced "higher quality" white flour than was true of the more labor-intensive techniques. This is a good example of a created local market demand where the highly milled and sifted flour is nutritionally inferior to that produced by other means.[12]

One of the few and most comprehensive discussions of the use of the bicycle as a fundamental tool in labor-intensive technologies has been written by Stewart S. Wilson.[13] He considers not only the use of pedal power as a means of transportation, but also the development of chain pumps, grinders, and other agricultural equipment powered by a stationary form of the same principle. The efficiency of simple gearing mechanisms powered by human legs is substantial. According to Wilson, a man on a bicycle ranks first among animals and machines in terms of transportation efficiency as measured by calories expended per gram-kilometer moved. In fact, the man on a bicycle expends only about one-fifth the energy of a man walking. (It is interesting to note in terms of contemporary technology that the jet transport also is somewhat more efficient, in terms of calories per gram-kilometer, than a man walking.)

Additional examples could fill several books. Thus far, most have rested on existing technologies. The factors that affect their implementation generally have been cost, ease of maintenance, reliability, and acceptance into the rural community. They often have been most successful because they can be accomplished with relatively low costs and represent tools that are locally maintainable with fundamental skills. Since they usually perform a single and independent function, reliability is a less acute factor, and a failure of short duration has minimal impact. As the tools are small-scale and of single purpose, they often do not represent a sizable intrusion into the local society or culture. In addition, they have the advantage of being highly demonstrable, which in turn leads to easier acceptance.

Urban Manufacturing

The next most researched category is the development of labor-intensive technologies for urban areas, almost always related to the production of consumer goods. This application needs continued expansion since the demographic reality in most countries is that urbanization will continue and that unemployment in the urban centers, especially the capital cities, is a persistent and expanding problem that can lead to the most adverse consequences. Some factors that affect the development and implementation of such technologies are costs, employee training and literacy, factory versus cottage concepts, quality control, and marketability. An extensive presentation on the manufacture of footwear in Ethiopia by M. S. McBain and J. Pickett is a representative example.[14] Agreeing with other authors, they conclude that an intermediate-sized establishment may be more effective, when both costs and number of people employed are considered, than either the major factory or the one-or-two-person cottage industry.

Data from Indonesia suggest that the range in the number of people employed per unit of output for consumer goods is substantial. For example, labor-intensive versus capital-intensive techniques might lead to employment ratios of 13:1 in the cigarette industry, 12:1 in the manufacture of flashlight batteries, and as high as 22:1 in the production of beverages and soft drinks. Even in the recapping of tires the ratio can approach 6:1.[15]

The David Livingston Institute of Overseas Development Studies has investigated the choice of technologies in a wide selection of urban industries, including leather manufacturing, iron foundries, milling, bolt and nut manufacturers, fertilizer, and African textile production.[16] Even the choice of techniques in the manufacture of cans has been studied intensively.[17] In summary, the development of various technologies emphasizing labor intensity has almost limitless applications.

Although not often cited in the literature, some emphasis also should be given to the development of crafts and craftsmanship in urban settings. The previously cited production of African textiles is an example. Opportunities for the manufacture of a wide range of products, some salable on the international market, should be expanded. The production of relatively high-quality, simple, contemporary wooden furniture is certainly possible. The preservation of indigenous craftsmanship and traditional motifs, be it in pottery, wood, textile, or other materials, could have an expanded local market and a possible international market. The fundamental barriers to such development are not technological, as most of the tools are readily available at modest capital investment, but tend to be related to management and marketing abilities.

Rural Public Works

The third category of labor-intensive technologies is public works and other construction projects that require the integration of multiple activities on a relatively large scale, employing many people. Most application would occur in rural areas. The construction of public infrastructure, such as water supply systems, sanitary systems, dams, or roads, are examples. Road construction is illustrative of the diverse difficulties encountered when using labor-intensive techniques in large projects. A transportation linkage, normally a road of some type, is a necessity for rural development. Without the physical interconnection of a transportation link, importation of needed commodities and services and export of surplus production are impossible. This need has been recognized by the development of a transportation technology support project in the Transportation Research Board of the National Academy of Sciences.[18] The International Labor Organization has published a substantial work on the construction of low-volume roads.[19]

Most large projects have emphasized the use of capital-intensive heavy equipment, imported from a limited number of developed countries. However, adequate roads were built long before the invention of such equipment. And not all roads need or should be of a high quality or paved. The Sudan, which instituted a road building program a few years ago, has recently completed the first paved road from Khartoum to the sea. It is clear that a country like the Sudan, with a million square miles, cannot afford either the cost or the time necessary to develop high-quality roads throughout its area. Roads need not be all-weather, and not all streams need to be bridged. Roads can be built in as broad a spectrum of design and quality as any other work of man.

What, then, are the barriers to labor-intensive techniques in road building? Many are the same as those identified earlier in this chapter, including the inadequacies of contemporary engineering education and the development of a very narrow engineering aesthetic. The costs and time of construction are always factors. However, they are factors that may be overemphasized in the development of many rural road systems. The performance of the system, which is exposed to the physical environment and to the abuses of an uncontrolled user group (overloaded lorries being a prime example), is an additional engineering concern. Most engineers consider inadequate performance of roads as a failure. The tendency, therefore, is to overdesign. Appropriate road standards and construction, and the local manufacture of road construction equipment, have been developed in Kenya through the Rural Access Roads Program.[20]

There is a myth that the higher quality the road, the more sophisticated the design, the greater the quality control (i.e., the more mechanized the equipment), the easier and cheaper the road will be to maintain. Over the last fifteen years there has been an expanding amount of research which indicates that although there is some truth to this idea in the narrow range, it is misleading in the broad range. In other words, a paved road is not necessarily cheaper to maintain or easier to maintain than a gravel road, nor a gravel road necessarily cheaper or easier to maintain than a dirt road.

In addition to the technical concerns, the use of labor-intensive techniques for large public works becomes a problem of management, training, and organization. Public works require many different skills and functions to create the final product. If labor-intensive techniques are used, these require large numbers of people. The training of these people, their organization, and management becomes a substantial problem. People with such training and management skills are in short supply in most countries. In addition, the developed countries have not provided the organizational structures or enough training tools to support such projects. More imagination is needed in these areas. I would suggest consideration of highly sophisticated tools to do this very unsophisticated job. The use of modern, easily communicated audiovisual training techniques may be highly effective where the trainees are illiterate. Their audiovisual capabilities may be more acute than those of a more literate, book-based society. In a similar vein, use of the simple, highly sophisticated hand calculator already has demonstrated its success.

Since a large number of people are involved, they probably represent a substantial portion of the total indigenous population in a given location. Therefore, the ability to institute and carry through such projects requires maximizing the use of the existing social and organizational systems. This calls for a substantial input from sociologists, cultural anthropologists, and others from the social sciences.

Conclusions

The intention of this chapter has been to alert those concerned with economic development to the potentials and problems in the concept of appropriate technology, especially the problems of instituting labor-intensive techniques. Clearly, there is a need for an integrative approach because of the multidisciplinary nature of the problems. This is especially true if large-scale, labor-intensive techniques are to be broadly applied. Although substantial work and much imagination are necessary inputs to develop new tools, the successful implementation of labor-intensive techniques is fundamentally a people problem. Therefore, training, management, social structure, and cultural systems become the four cornerstones upon which success of technical activities must rest. The failure of any one will either substantially impede the implementation of the technology or defeat its chances for success.

Finally, the emphasis on labor-intensive techniques in this chapter reflects the fundamental belief that the most valuable resource of any country is its people. The development of this resource cannot be instantaneous. It requires time and patience. More important, however, it requires opportunity and resources. The use of capital-intensive technologies has led to the development of dualistic economies in many nations. Some would say this is wrong. They feel that development should be totally centered on rural development to raise the quality of life for all of the population simultaneously. A more balanced view would seem to be supported by the appropriate technology approach. Modern Third World countries wish to become just that—*modern* Third World countries.

Appropriate technological analysis would indicate that there is a place for the most advanced twenty-first century technologies in most countries. There is a place for modern capital-intensive production. There also is a place for labor-intensive activities. This chapter has identified some of the real barriers, both internal and external, technical and social, to greater and more accelerated application of labor-intensive techniques. Greater efforts by all concerned must be made to encourage and implement these techniques.

2. Intermediate Technology and Development

Warren Adams

Since World War II there has been great concern about the need to promote more rapid development within the countries and areas of the Third World. Development programs have been supported by bilateral and multilateral assistance from official and nonofficial entities. The objective was development, but the emphasis of programs periodically shifted from development to land reform, community development, Five Year industrial plans, export promotion, import substitution, Green Revolution, family planning, and others. Time has brought greater appreciation of the complexities of development work, and the historic changes in direction should caution us against accepting a single correct thesis or hoping for a general panacea.

Assessments have been made of the results of development efforts. The usual economic indicators, i.e., GNP, national income, food production, have shown reasonable progress, but rising populations have made per capita measures less positive. Other indicators have demonstrated reasons for less than satisfaction. Unemployment and underemployment refuse to melt away, and even increase; balances of payments deteriorate, and indebtedness rises; income disparities continue to grow between developed and developing countries and within developing countries. Some newly emerging countries are dismayed to find their newly acquired political freedom unmatched by economic independence. There is a tendency to ascribe these problems to the failure of the "system" to balance and to deliver the promised benefits of development.

The Concept of Intermediate Technology

The critical role of technology in economic development, and especially the importance of technological choice, was first brought into focus by the late E. F. Schumacher nearly twenty years ago.[1] Such choices influence not only production processes but products as well, the location and conditions where man works and lives, and man's relation to his environment. Yet, in truth, decision makers in developing countries appear to face a Hobson's choice: stay where you are with your traditional technologies, or adopt the technologies of the industrialized countries and struggle to adapt to them.

It has been assumed by many that technologies from industrialized countries were appropriate. They are validated by success, they have status in the eyes of educated leaders in both developed and developing countries, they are "bankable," easily located, and readily available. Developed for Western conditions, these technologies typically are capital intensive (labor saving), large scale, and heavily centralized. They require for their efficient operation an elaborate support structure of markets, training, raw materials, supplies, spare parts, and cheap, efficient communications systems.

Schumacher, despite his reputation in orthodox statistical and monetary economics and his job as economic adviser to the British Coal Board, was quick to observe two basic conditions in developing countries: (1) The bulk of the populations are rural and live in a poverty trap characterized by high unemployment and serious underemployment. (2) As a result of the above, there is a virtually inexhaustible potential for mass migration from the rural areas to the cities, many of which already face insoluble social, economic, and environmental problems.[2]

Unfortunately, large-scale, capital-intensive technologies aggravate those problems. Being urban-centered, they are in the wrong place; the kind of goods they produce are generally of the wrong kind to meet the needs of the poor; and being capital intensive and labor saving they are unable to provide sufficient jobs to cope with the growing supply of labor. It is, of course, not only employment that is the problem: capital-intensive technologies also make demands upon special types of infrastructural facilities, shape educational standards and norms, influence consumption patterns and life-styles, and dictate import-export patterns. Technology is neither economically nor culturally neutral; it never has been and never will be.

The essence of the argument for intermediate technology (IT) is therefore that the capital-intensive, centralized, complex, and costly technologies of the rich countries are generally inappropriate for poor countries, and especially their rural communities. To meet *their* needs, a new technology must be discovered or devised that lies between the traditional and ultra-modern, or, so to speak, between the sickle and combine harvester.

Schumacher named such technologies "intermediate" primarily because their cost per workplace would lie somewhere between the £10,000 it costs to equip one workplace in a typical rich country, and the negligible cost of traditional tools of a rural peasant or artisan in a poor country. He consciously rejected using the term "appropriate technology" to convey his meaning because he felt it was a "question-raising" expression: appropriate to whom? to what? and where? The importance of locating intermediate technologies is to provide technological choices so people can answer for themselves what is appropriate from a range of alternatives.

However, in insisting on technological choice we must make a clear distinction between science, on the one hand, and technology, on the other: between scientific knowledge and its applications. The knowledge of scientific laws, of materials and of methods, is, in a sense, absolute, and one could hardly talk of intermediate knowledge or intermediate science. But the application of the best knowledge can take many different forms and can lead to many different types of technology and modes of operation. It is here that the need for and the possibility of intelligent choice enters.

Different economic and social conditions demand different applications. It is therefore a complete misunderstanding and often a wilful misrepresentation when people accuse the proponents of intermediate technology development of wishing to withhold sophisticated, capital-intensive, labor-saving "advanced technologies" from poor societies or of offering intermediate technology as the only mode of production to be considered. None would deny that there are conditions in which the most sophisticated technology is the most appropriate, and other conditions in which an intermediate technology is most apt. However, as long as an intermediate technology does not exist or is inaccessible because of a lack of knowledge and communication, the people in the latter situation have no useful choice. Either they do nothing at all, or they do the wrong thing by trying to use an inappropriate technology with usually negative results.

Intermediate technologies fall between the one extreme of advanced Western technology, and the other extreme of the traditional method of doing things. An intermediate technology often will be more efficient economically than either of the other alternatives because it will produce more output from given expenditures on labor, capital, and other inputs. It therefore will have lower unit production costs. It generally will involve a technique more complex, and requiring more capital, than traditional methods, but will be simpler and less capital intensive than the Western technology. In many instances around the world it has been found that an intermediate technology is more appropriate (in many senses) to the conditions of developing countries than the Western alternative, primarily because it avoids the problems of Western technologies mentioned above, and yet provides an economically viable (i.e., profitable) way of doing things.

An intermediate technology generally has the following features:

(1) It is small scale so that it can be replicated throughout the country, located near its market and its source of raw materials, and provide employment where the people live.

(2) It involves low capital costs so that it can be set up within the country's capacity for saving for investment in productive enterprise, and so that a plant is not beyond the ability of small, indigenously owned and run businesses and cooperatives to acquire the necessary capital.

(3) It is labor intensive (or capital saving) so that it conserves scarce capital and creates many jobs.

(4) It uses simple techniques so that they are within the capacity of the local people, given their educational facilities, to master and develop for themselves.

(5) It relies on local resources as far as possible so that the country's natural resources are developed, the initial investment has a multiplier effect on local employment, and the country can escape financial and physical dependence on erratic and scarce imports.

(6) It is profitable so that it creates a surplus that is the source of the country's

future income and investment, and so that entrepreneurs (or cooperatives, or whatever social arrangement the country has for organizing productive activity) have an incentive to undertake the investment.

These features go against the conventional trend of technological and organizational development, which is toward units of ever larger scale. This approach is said to be justified by the economies of scale. Large-scale production units, however, tend to create many sociological, ecological, and resource problems, the burden of which normally has to be carried by the community at large and does not enter into the unit's cost calculation. Even from a narrow economic point of view, large units are economical only when certain conditions are satisfied: high market density and/or highly efficient, reliable, low-cost transport systems, skill in large-scale administration, management, buying, selling, and so on. When these conditions are not satisfied, so-called economics of scale become illusory. In any case, large scale tends to act as a principle of exclusion: only people who already are rich and powerful can embark on new productive enterprises. The man of small means is excluded and reduced to the position of a job seeker, and when there are not enough jobs provided by the rich and powerful, he has no reasonable possibility of becoming productive. Smallness is, in fact, a precondition for rural development, and it is now becoming increasingly relevant from social, ecological, and resource points of view.

Much the same applies to simplicity and capital saving. It does not take great technological creativity to take a further step in the direction of complexity, capital intensity, and giantism. The suggestion to engage the best of modern knowledge and intelligence in the search for smallness, simplicity, and capital saving almost invariably meets the argument in the first instance that it cannot be done, or, if done, it would prove to be totally uneconomic. In this matter prejudices and untested presuppositions are very deeply rooted. There is now accumulating evidence that it can be done, but it requires a more creative and original research development effort than is normally forthcoming.

It has become increasingly obvious that as far as employment, and especially rural employment, are concerned, the application of these principles is the only feasible course of action. If developing countries are not offered a range of choices in technology, or if they are persuaded that no such choices exist or can be created, then there is no hope for the mass of rural and urban poor. This can be illustrated by the following example, used by Schumacher:

> According to my estimates, there is in India an immediate need for something like 50 million jobs though others put it as high as 80 million. If we agree that people cannot do productive work unless they have some capital—in the form of equipment—and also some working capital, how much can we afford for each job? Now, if it costs £10 to establish a job, you need £500 million for 50 million jobs. If it costs £100 to establish a job, you need £5,000 million. And if it costs £5,000 per job, which is what it would cost, say, in Britain, to set up 50 million jobs, you require £250,000 million. Now, the national

income of the country we are talking about—India—is about £15,000 million a year, so that first question is how much can we afford for each job, and the second question is how much time we have to do it in. Let us say we want 50 million jobs in ten years. What proportion of national income (which I identified as about 15,000 million) can one reasonably expect to be available for the establishment of this capital fund for job creation? I would say, without going into any details, you are lucky if you can make it 5%. Therefore, if you have 5% of £15,000 million for ten years, you have a total of £7,500 million for the estalishment of jobs. If you want 50 million jobs in those ten years, you have a total of £7,500 million for the establishment of jobs. If you want 50 million jobs in those ten years, you can afford to spend an average of £150 per workplace at that level of capital investment per workplace; in other words, you could afford to set up 5 million workplaces a year. Let us assume, however, that you say: "No, £150 is too mean. It will not buy more than a set of tools; we want £1,500 per workplace." Then you cannot have 5 million new jobs in a year, but only half a million. And if you say, "Only the best is good enough. We want all to be little Americas right away, that means 5,000 per workplace," then you cannot have half a million new jobs a year, let alone 5 million, but only about 170,000. Now, you have, no doubt, noticed already that I have simplified this matter very much because in ten years with investment in jobs, you would have an increase in the national income, but I have left out the increase in the population, and I would suggest that these two factors cancel one another in their effect on my calculation.[3]

Organization of the Intermediate Technology Development Group

The virtual absence of detailed, practical information of intermediate technologies led Schumacher and a group of friends to set up the Intermediate Technology Development Group (ITDG) in 1965. The Group's aim was, and is, to help fill this "knowledge gap" and thereby provide more effective technological choices.

The Group took its first positive action with small donations. A short catalog of animal and man-operated agriculture equipment still available from small firms in Britain was compiled, mimeographed, and sent with an aid mission to Nigeria. Response was so encouraging that in 1967 a larger catalog, *Tools for Progress: A Guide to Small-scale Equipment for Rural Development*, was produced; seven thousand copies were sold, and the money borrowed to print it was repaid. The success of this catalog has led to subsequent publication of modified and expanded versions of *Tools for Progress*—work is beginning on a fourth version now—and laid the groundwork for a major sector of Group activity in publications. By the early 1970s publications were spun off to a subsidiary, IT Publications Ltd., which began to publish a quarterly journal, *Appropriate Technology*, as well as a growing range of technical studies, designs and drawings, bibliographies, and more.

Respondents to early publications often commented on modifications to technologies that had been helpful under different situations, or raised questions about difficulties experienced, or suggested further areas of need. The Group quickly became aware of the need to develop wider scientific and technical capacities to deal with the range of technical inquiries flowing in from overseas. A group of friends, who were scientists and technicians, formed a panel to help answer the technical inquiries and to advise the Group on its work.

The idea of using professional people on voluntary panels soon became a distinct mode of ITDG operations. The early panels started with building materials and methods, water, agricultural tools, and power; they now have expanded to nearly two dozen subjects. Panel members share two characteristics: they are highly competent and experienced in their field, and they are interested in exploring the role for intermediate technology. The panels typically set these four goals as their objectives: the use of technical inquiries and other field observations to establish the basic needs requiring assistance; the ascertainment of what is known and what gaps exist in meeting those needs; the actual efforts to fill the gaps; and then the dissemination of information about the technologies. Panel members volunteer as individuals and are drawn from academic and research facilities, businesses, and public establishments. Thus, the present panel membership provides the Group with over three hundred links to research, training, and action agencies that constitute an invaluable feedback resource linking together needs from the field, ongoing research, technical capacities, and wider professional capacities, which can be mobilized as needed.

The panels have provided the basis for many of the Group's publications over the years. However, as the reputation of the Group has grown, and as intermediate technology (IT) has been increasingly accepted, panel members have helped to meet the requests for consultants on special problems on which the IT viewpoint is requested. The subject and geographic range of such consultancies is large and growing, and the range of funding support for these activities is another measure of acceptance. While some developing countries are able or prefer to fund a consultancy themselves, other cannot and seek funds from a range of aid agencies—national or international, private or official—to obtain assistance. In some instances we are able to help locate funding or to deal with the need in conjunction with other field activities. As with publication activities, consultancies are handled by a subsidiary established in 1969 with the purpose of keeping ITDG's administration decentralized and the subsidiary self-supporting.

In the early years the interest of the panel and consultancy subsidiary led to projects in four countries—Zambia, Nigeria, Jamaica, and Tanzania—where staff were located to work on agriculture, rural workshops, and small-scale water supply and food technology. Additional technical staff were added slowly in selected areas, such as power, agriculture, and water, to work more directly with the panels in these fields. Over the years staff for the field projects have declined in number as the focus of Group efforts has shifted from administration of field projects to shorter term expert consultancies, answering technical inquiries, and providing basic capacities to act as a catalyst and to "know who knows what" so as to be able to mobilize responses quickly. Technical staff are located out of London near facilities suitable for their work needs; thus a small cluster of staff working on agriculture, water, and power are located at a campus facility of Reading University, where they can relate to each other and be involved with university work and facilities as well.

Technical staff, while continuing to relate to the panels, are increasingly active on consultancies. The power officer is working on a World Bank-funded project to identify small-scale solar pumping units. The stove officer, funded by the British Hildon Trust and Overseas Development Administration, is doing a consultancy program with groups in Third World countries. A project officer working on river current turbines is funded by the Dutch government to construct and test a full-scale model in South Sudan, etc. For some years, ITDG seconded two persons as staff to the UN's Economic Commission for Africa in Addis Ababa, but they have now returned to the Group. Currently the Group has a single resident representative to Africa, who operates from Nairobi to keep in touch with appropriate technology developments in African countries. Finally, there is a field project in Juba, Sudan, where the Group has been successful in launching a ferro-cement boat building program, training Sudanese staff, and is currently turning the project over to other agencies.

One of the examples mentioned in *Small Is Beautiful* concerns the need to develop an intermediate technology to manufacture egg cartons for Zambia. A subsidiary, Development Techniques Ltd., was set up for this purpose and ultimately proved highly successful. A range of different-sized egg carton machines have been designed, produced, and are now in operation in over a dozen countries. Manufacture of this equipment is now under license to a British company, but the machines continue to meet a variety of needs, and additional, useful modifications are being pursued by the Group.

The Group came to realize that IT has application for other than rural areas. The needs of the informal sector in urban areas are closely related to rural populations, and backward and forward linkages between conventional industries and IT can be established both in rural sectors and the informal sectors. Thus, when the British government elicited the Group's ideas about what further IT programs it might usefully perform, ITDG was able to suggest work in the field of industrial technologies.

For the past few years, with funding from the Overseas Development Administration, the Group has operated a unit called Intermediate Technology Industrial Services (ITIS), which is located in Rugby, England, and predominantly staffed by engineers. In the industrial sector this unit's functions are to answer inquiries about existing technologies for specific small-scale industrial activities; recommend sources of supply for machinery and equipment; identify the requirements for new products or processes through field visits by technical experts and market studies; assist with the adaptation of existing technologies and the development of new technologies; and provide funds to field-test and demonstrate new technologies in developing countries. Because their activities are funded, the engineers can and do perform a more active role in traveling through their areas to explore technical inquiries and seek out companies (at home and abroad) that have the most relevant equipment or know-how. They help overseas clients select the most appropriate technologies and facilitate direct exchanges between inquirers and the chosen manufacturers.

ITIS is currently financing projects in various countries by feasibility studies and market surveys; development of new technologies; field-testing existing and new technologies; and pilot industrial projects. Also, although the funding for this unit was recognized to be "risk" capital in major part, and large expenditures require approval of a joint committee, the unit seeks local contributions from its collaborators, be they governments, private institutions, or individuals.

Thus, for example, in addition to ITIS funds and expertise, local funds for a wool-spinning project in India are supplied by the government of Uttar Pradesh State, or in Sri Lanka, a private entrepreneur and local commercial bank are investing in installation of a small-scale glass plant. Further, the unit places great emphasis on working through local intermediaries who have a commitment to the project; the intermediary may typically be a government agency, a university or research group, or more commonly a private organization or individual, often a nonprofit charitable association. The ITIS staff have answered nearly six hundred technical inquiries per year, done or commissioned several dozen feasibility studies and market surveys, financed the publication of over twenty books and catalogs on appropriate products and processes, and has about seventy projects under way. Although the initial three-year funding has concluded, there are good expectations that this rather special example of collaboration between government agencies will be continued.

At the same time that the British government decided to fund the formation of the ITIS unit, it also agreed to set up a fund for ITDG by which encouragement could be given to development of appropriate technology institutions. Since its inception, ITDG has pursued a policy of locating, encouraging, developing, or doing whatever possible to build on local resources. Over the years ITDG has developed and maintained contacts with a number of groups with similar interests in other industrial countries; knowledge of special interests in other capacities has made it possible to refer inquiries or requests for assistance more quickly to others capable of taking effective action. More significantly, ITDG has been developing an even closer linkage with action groups in developing countries. In some countries, such as India, the tradition of self-help or cottage industries is strong, so related activities have long been present, and ITDG simply seeks to work with such groups to strengthen them where needed, collaborate with them on projects of mutual interest, and help to link them into the larger network of groups working on IT. In other countries the idea is new, and there is little local experience to work with; but individuals and small groups once in place can begin to develop indigenous capacities to identify needs, locate resources, and assist in the vital tasks of modification and adaptation of IT locally. IT groups are called by a variety of names, and they operate from substantially different positions: some are government agencies, others are "parastatal," academic-based, charitable, nonprofit, etc. The important consequences of the growth of the network is that local self-reliance is enhanced, and internationally the capacity to be of mutual assistance is qualitatively greater.

Therefore, the provision of an "AT Institutions Fund" by the British Overseas Development Administration has significantly enabled ITDG to pursue additional ways of building local capacities for self-help in developing IT. These funds are not available for capital expenditures but can be used for a wide variety of training and staff upgrading to benefit Third World IT institutions. In Africa staff have traveled from Gambia and Upper Volta to the Technology Consultancy Centre in Kumasi, Ghana; a senior staff member from the center came to England for further specialized training; and his director went to the Appropriate Technology Development Association in Lucknow, Uttar Pradesh, for a six-month collaboration. Regional meetings of IT institutional staffs have been held in Peru, Chile, the Philippines, and Sierra Leone. These developments are both highly important to and valued by ITDG because they are, quite appropriately, moving the activity closer and closer to the source of problems. In the interim they also insure that current programs benefit from increased participation and contributions by local activists.

Intermediate Technologies in Practice

There is now a growing literature of case studies of appropriate technologies under actual operating conditions.[4] The following examples illustrate what has been and is being done by way of modifying and adapting technologies in a number of industries where the intermediate level of technology was formerly unknown or did not exist.

Transportation

In the transportation field the value and use of IT increasingly has come to be accepted. Members of the Group have been associated closely with the early innovative work of the ILO, the Kenya Rural Access Roads Program, and served as consultants to the World Bank and various national governments. Work in the field has included road and water transport, tools and equipment for construction, and simple vehicles.[5]

The early work of ITDG was largely through panel responses to technical inquiries and requests for consultants. However, the addition of a technical officer in 1976 led the panel to produce a series of Information Papers about its work and ideas; helpful responses were received from technical specialists all over the world, and this led to still more work in other parts of the world, contributed to further improvements, and led to additional work opportunities for the panel.

In 1979 the panel chairman and project officer formally incorporated, as a subsidiary company, IT Transport Ltd. and launched operations with funding assistance from Christian Aid. The major purpose of the company is to provide professional advice and assistance in the assessment, development, and implementation of appropriate transport policies and technologies for developing countries (initially) and for the United Kingdom (subsequently). The primary means for achieving these objectives is through consultancies, and the first year of activity has established a viable footing with encouraging prospects. Most recently, the following major assignments have been carried out for several sponsoring organizations:

Investigation of the share-taxi industry in Nairobi to identify the need for an appropriate type of assistance to overcome operating problems and improve service (World Bank).

Assessment of the quality and availability of tools and equipment suitable for labor-based road construction and maintenance in Botswana, specification of the implements to be used on the pilot project, and advice on policies leading to local manufacture (International Labor Office).

Advice on the design and manufacture of cycle trailers, load-carrying tricycles, and low-cost motorized vehicles to an established major industrial undertaking in Sri Lanka (Intermediate Technology Industrial Services/ODA).

Investigation of the feasibility and design of animal-drawn transport for Western Samoa. Appraisal of the availability of local materials and expertise for the construction of suitable vehicles and harnesses (Economic and Social Commission for Asia and the Pacific).

Supervision of a program of overseas testing of the Oxtrike (an improved tricycle) in several Asian and African countries to establish the suitability of the vehicle for use in developing countries, and to evaluate its market potential prior to initiating local manufacture (Intermediate Technology Industrial Services/ODA).

Development of a methodology for costing road accidents in developing countries and valuing accident prevention measures (World Bank/University of Newcastle upon Tyne, U.K.).

Study of the effect of rural road program on poverty alleviation and measures to make investment more effective in helping the poor (International Labor Office).

Work on the Oxtrike project and advice on the Bangladesh Planning Commission and International Labor Office continued into 1981. However, it has become increasingly clear that the lack of appropriate basic vehicles is a major constraint on the activities of many developing communities. To participate in efforts to overcome this problem, IT Transport is establishing a development workshop. This will enable suitable prototype evaluation to be undertaken prior to field testing and manufacture in developing countries. The workshop became fully operational early in 1981. Proposals have been drawn up for the development of a Small Farm Transport Vehicle, cycle trailers and—in collaboration with the International Institute of Tropical Agriculture—a simple, motorized vehicle to complement the no-till system of agriculture.

Buildings and Building Materials

One of the earliest of ITDG's panels was formed in 1968 on the subject of buildings and building materials. The first project of this panel was a request for assistance in upgrading the efficiency of small, local building contractors in an African state.

In responding, the panel argued that as a critical first step these small entrepreneurs needed to develop basic managerial skills. Without such skills, no amount of other technological developments would be of much value to them; in other words, they initally needed "soft" rather than "hard" technology. The panel realized that small builders were outside the range of most training programs, and that this represented a significant gap in the construction industry, which is an important bridge between agriculture and industry and a major contribution to gross capital formation and employment in most countries. The problem, typical to many African countries, was how to develop the construction industry, especially in the rural sector, without the indigenous skills to execute such programs.

Our "Building for Development" team, made up of an architect, an engineer, and an economist, carried out fieldwork in Nigeria and Kenya. They experimented with ways of presenting management training material to contractors through lectures, seminars, films, exhibitions, and dramatizations. Having developed and demonstrated a capacity to impart managerial skills to contractors—most of them men with little formal education—the team was invited to introduce these field-tested training techniques into courses run by the National Construction Corporation of Kenya. They produced a series of teaching kits that the Group published in the early 1970s.

The engineer of this team, Derek Miles, remained with the Group as a consultant. In 1976 he helped the International Labor Office organize an African regional construction management course held for eighteen nations in Nairobi. Two years later, and eight years after the building team had shown the way, the ILO launched a thirty-month construction management training program in which twelve African countries participated. Development of a parallel set of training courses on project management for Ministry of Works staff of African countries was instituted by the ILO with the Group's involvement. Derek Miles has written other useful manuals and served as consultant to numerous national training programs in Africa and Asia for the ILO, UNIDO, World Bank, UNESCO, and Norad.[6]

Highly efficient, small-scale brickworks have been designed and are in operation in several African countries. A typical modern brick factory in an industrialized country produces over a million bricks a week. The mini-brickworks, which is either hand-operated or uses very simple machinery, produces some ten thousand bricks a day. Capital costs per workplace in the small unit are about £400 as against some £40,000 in a large, modern brickworks. In small units, very large savings on fuel costs are possible by air-drying the bricks before firing, and local production virtually eliminates transport costs. Such brickworks established by ITDG consultants during the past few years now create employment for over a thousand people. The IT unit responsible for developing and installing mini-brickworks also has developed an on-site method of making reinforced corrugated roofing sheets using cement and a variety of fibers. These cost an average of about one-quarter of imported roofing sheets.[7]

Mini-cement plants in China, where 50 percent of the total output of 39 million tons of cement is made in small-scale vertical shaft kilns, recently have been described in some detail.[8] A considerable amount of work has been done in India on such mini-cement plants, and the Appropriate Technology Development Association, Lucknow, is in the course of implementing a pilot commercial project for a vertical shaft kiln making portland cement with a capacity of twenty-five tons per day. The feasibility study indicates a capital cost of about half that of the large-scale technology per unit of production capacity; lower scale requirements for installation and much more rapid installation (about a quarter of the gestation period required for a large-scale plant); lower transport and distribution costs; lower quality of fuel required; and, in addition to the wide dispersal of small scale in rural areas, the use of small deposits of calcarious materials that otherwise remain unused.

Alternatives to cement, based on lime, brick dust, and other local raw materials, also have been developed in India. This mortar is suitable for many kinds of construction required in rural areas. Development work is still proceeding, but it is already certain that this alternative to cement can be produced cheaply and in small quantities. A pilot plant has been erected in Tanzania with a capacity of three tons per day for making building blocks, mortar, and plaster. Additional plants are being considered.

Power

There is now widespread public interest in renewable energy resources. However, well before this development, the Group formed a power panel and appointed a full-time power project officer. Together they responded to technical inquiries, collecting and disseminating information on small power systems relevant to the needs of developing countries. From the early interest in wind, the Group developed interest in other resources, e.g., water, solar, and steam, among others. New panels followed, and a power program was formed to deal with the following:

(1) Wind. One of the earliest efforts of the power panel was in the area of windpower for pumping water. The project officer examined windmill technologies in the Omo River area of Ethiopia.[9] He then began to work on development of a windpump system suitable for small-scale local production, relying on commonly available steel and adaptable to modifications meeting the requirements of high-volume, low-head irrigation pumping or low-volume, high-head pumping of drinking water for people or cattle.

Following a two-year prototype development and testing program at the University of Reading, funding was obtained for an extensive world testing and manufacturing program with collaborators in several countries: Kenya, India, Pakistan, Egypt, Oman, and Botswana. The record of production, modification, and sales in Kenya are encouraging, and the outlook in India and Pakistan is good. Concurrent instrumentation and testing of wind generators also is being undertaken in Kenya.

(2) Solar. One of the concerns of the solar panel was related to passive solar heating. Through the use of ITIS funds, a panel member launched a project to introduce residential passive solar space-heating systems in the Ladakh region of northwest India.

Also responding to a request from the World Bank, the Group joined in collaboration with William Halcrow and Partners in a long-term project to test and evaluate currently available technology for using solar energy in small-scale irrigation pumping under both laboratory and field conditions. The relative merits of photovoltaic and thermal systems have been considered in the first two years, and suggestions for future development, especially for local manufacture, are under consideration.

(3) Other. Testing of a vertical-axis, cross-flow turbine to extract energy from slow-moving rivers for water pumping and electricity generation has been completed in Juba, Sudan. A novel electronic control system for small-scale hydroelectric plants is being tested; it costs less than conventional mechanical, flow-control systems and promises greater reliability and less maintenance. Additional programs are in hand for design, testing, and implementation of improved cooking stoves, steam engines, and the Humphrey Pump.

Intermediate Technology in Developed Countries

The original focus of the Group's work was on the needs and problems of technologies in the developing countries. However, in more recent years requests for advice and assistance have been received from groups and individuals in developed countries. Academics have sought ways to bring IT into their curricula; business firms, workers, and distressed communities contemplating increased unemployment have sought useful alternative opportunities; and individuals have expressed concern about their environment.

Fortunately, there have been signs of progress on these problems, and the Group has been a participant in some of the steps being taken. More academic and research institutions have been introducing lectures, courses, and degree programs relating to the work and issues of appropriate technologies. Several major professional institutions and associations in Great Britain have held conferences to discuss and explore IT and their profession, seeking ways to develop this concern into a more active role for the institution. George McRobie devotes several chapters and gives lists in several appendixes of his book, *Small Is Possible*, to describing and identifying the long list of groups, academic institutions, and communities active in work relating to IT.[10]

In 1975 John Davis joined the Group to work on Appropriate Technology for the United Kingdom (AT-UK) issues and programs. A collection of his lectures describes the early efforts of the Group in Great Britain.[11] In recent years, as needs have been more fully articulated, the work of the AT-UK unit has tended to focus on four related activities: Local Enterprise Trusts (LETs), Local Energy Groups (LEGs), a Small Firms Research unit, and a Technology Choice project.

The aim of the LETs is to stimulate and assist new local growth and to reduce risks and prevent unnecessary failures in the vulnerable early years of new enterprises. Some fifteen LETs are actively operating and demonstrating that they are a cost-effective, self-help way of fostering new enterprise, generating employment, and encouraging local self-reliance. More communities have evidenced interest in LETs, and there is a growing public response to the basic four R's that underlie many LETs: repair, recondition, renovate, and recycle. Some of that same interest is reflected by LEGs, which are forming themselves in various parts of the United Kingdom.

The Small Firms Research Unit has been examining ways and means of facilitating changes in the environment within which small firms operate. Similarly, the Technology Choice project has identified a number of sectors for study, including food and agriculture, health, housing, energy, and transport. Food and agriculture have been selected as the initial target.

Conclusions

After more than fifteen years of experience the Group is satisfied that there is indeed a need for IT and that a substantial knowledge gap about IT still exists. While the early days of the Group's life were devoted to proselytizing, especially while Schumacher was alive and lecturing, the focus has shifted increasingly to locating, testing, and disseminating information about intermediate technologies. More recently the Group has focused its energies on institutional problems of implementing IT in developing countries and encouraging local measures that would support, maintain, and replicate these intermediate technologies. The different phases of the Group's activities are cumulative, not successive. There are still many who are unconvinced, the knowledge gap is great, and the institutional problems are elusive and require constant attention.

3. The Political Economy of Intermediate and Appropriate Technology

Pat McGowan

> When Fritz Schumacher began to undertake such commissions to advise developing countries he felt a great sense of loneliness on his return. Most of his economist friends laughed at his low-cost, small-scale development ideas. Governments of the countries he advised paid only lip service to his counsels. The industrialists, and most official development agencies, were interested only in selling the latest and most sophisticated hardware to the poor countries.
>
> Verena Schumacher, *Small Is Possible*

One need not share the late E. F. Schumacher's vision to appreciate the potential of intermediate and appropriate technology in the development process.[1] As the chapters in this book illustrate, technologies based on the principles of smallness, simplicity, capital saving, and nonviolence not only exist but can dramatically enhance employment and rural incomes in the less developed countries (LDCs).[2] Indeed, the record of appropriate and intermediate technology in contributing to people-centered economic development is already so impressive that one is driven to ask why such technologies are not more widespread among LDCs and in economically depressed areas of rich industrial countries such as the Appalachia region of the United States.[3] What features of the political and economic organization of our world inhibit the greater use of intermediate and appropriate technologies? Why, in fact, has there developed over the last twenty years a movement aiming to foster the creation and use of intermediate technology? Which sociopolitical forces favor this movement and which (consciously or unconsciously) work against it? In sum, why has intermediate and appropriate technology surfaced as an international issue in such forums as the United Nations and among scholars and practitioners concerned with the development process?[4]

These are very big questions that do not lend themselves to simple, easily discovered answers. Yet answers must be sought, even if they are tentative and imperfect, because of the fundamental importance of the issues addressed by such questions. The future of some three billion people may depend to a great extent on the application of appropriate technology to meet their needs for jobs where they live and for an adequate quality of life. This chapter will offer some possible answers to these questions by focusing on the operation of what I shall call the modern world system (MWS), which is but a shorthand means of referring to the international political economy of modern times.

In searching for answers my approach will be historical and analytical. Historically, it easily can be shown that technological invention and innovation have been key forces driving the five-hundred-year history of the MWS. Analytically, it also can be shown that control of technology—and the potential to deny others access to technology—is a focus of class struggle and the basis of economic and political power in the modern world. Few would deny these points. But what is not equally well recognized is that within the MWS the continual pursuit of power and profit by states and firms is biased in favor of the most advanced technologies. As a consequence, I will argue the political-economic organization of our world is biased against intermediate and appropriate technology. The more widespread use of these technologies, and the achievement of Schumacher's vision of "economics as if people mattered," may depend on a fundamental restructuring of the world's political and economic organizations.

Basic Elements of the Modern World Systems

> The most obvious facts are the most easily forgotten.
>
> R. H. Tawney, *Religion and the Rise of Capitalism*

The MWS can be described from a static view of its contemporary structures or from a dynamic view of its historical development. The first approach gives a snapshot of the system's primary elements, whereas the second provides a moving picture of the laws of motion of the system. Both types of description are essential, but a static view of structural elements is perhaps a less complex and therefore a more appropriate starting point. What, then, are the structural constants of the MWS?

Immanuel Wallerstein, who is an American sociologist much influenced by the works of Marx and the great French historian Fernand Braudel, has argued that there are three structural constants that have characterized the MWS since its origins in the long sixteenth century from 1450 to 1650.[5] They are (1) a worldwide, hierarchical division of labor based upon long-distance trade in basic commodities, (2) a mode of production or economic system featuring capitalist commodity production for profit in the world market, and (3) a state system in which political and cultural pluralism exist so that the world's peoples cluster in different sovereign states and nations, some of which are culturally homogeneous while others are culturally plural.[6] How each of these basic structures affects the choice of technology deserves to be examined in more detail.

Trade and the International Division of Labor

As Europe began the transition from feudalism to capitalism sometime after 1450, the revival of long-distance trade over land and by sea began to link European cities with each other and the Mediterranean basin with the North Sea–Baltic Sea basin.[7] This trade produced increased specialization as cities and regions began to produce the commodities that their factor endowments (resources of capital, land, labor, and technology) and cost structures made most competitive in the international market. After the 1520s, with the incorporation of Mexico and Peru into the system, this trade network and its division of labor linked the New and Old Worlds, and, hence, was increasingly modern. The mere existence of trade over extremely long distances was not the distinguishing feature of the early MWS, however. Rather, the essential feature was the emerging hierarchical division of labor that was in the process of creating distinct types of political economies, which we may call "core" and "periphery."

By the 1600s a core had developed in northwest Europe (England, the Low Countries, northern France) that increasingly specialized in the export of commodities produced by relatively high-wage labor via technologies that, for their time, were capital-intensive and productive. Key core export commodities were textiles, ships, arms, and processed foods such as cheese and salted fish. This core area imported gold and silver bullion, wheat and other agricultural products, and raw materials including timber and naval stores. These basically unprocessed commodities came from the system's periphery: Spanish America for gold and silver, and the Baltic region, particularly Poland, for wheat and other grains.

Thus, there was a division of labor in which one region (core) engaged in manufacturing and exchanged manufactured commodities for foodstuffs and raw materials produced in another area (periphery). This is a familiar enough pattern today. For if in the sixteenth century core ships and textiles were exchanged for peripheral wheat and lumber, today core computers, aircraft, and industrial robots are exchanged for peripheral textiles, oil, and minerals as well as basic commodities and raw materials. *But what distinguishes the core from the periphery and vice versa is not the commodities exchanged per se, but rather how they are produced and the differing proportions of labor and capital that are employed.*

In sixteenth-century Poland and in the LDCs today the commodities produced and traded internationally result from a technologically labor-intensive process in which workers receive low wages and use little capital.[8] Core commodities, be they industrial or agricultural, are produced by high-wage labor using machines and processes involving advanced technology. This high capital intensity was as true of the core Dutch shipbuilding industry around 1650 as it is true of the core Japanese shipbuilding and automobile industries of the 1980s.

For those concerned with appropriate technology, two things are important about this core/periphery division of labor. First, while the actual geographical areas change over long periods of time, the basic structure of this division of labor is remarkably stable and has been a feature of the MWS for nearly five hundred years. It is not likely to be changed easily by anti-systemic forces such as the appropriate technology movement. Only if states and firms, as the actors with real political and economic power, act in concert will such a deeply entrenched structure change.

Second, specialization in response to factor endowments, market forces, and political action creates very distinct national economies: some rich, some poor, some developed, others not, all coexisting within a single MWS. These different economies are sharply different in their use of technology, largely because of core interests. Core economies are technological innovators and exporters of advanced technologies. Peripheral economies tend *not* to produce technologies appropriate for their needs. Rather, core countries sell to the periphery those advanced technologies that core economic and political interests make profitable. These technologies are located in LDC cities, while in the rural areas where most of the people still live the technologies used are traditional and more often than not of very low productivity. In the perspective of international political economy these different patterns of technology use appear to be a consequence of historically conditioned specialization and participation in the world economy. There is a division of labor and long-distance trade in technology just as there is for most other basic commodities, and the differences between core and periphery show up here perhaps more clearly than with any other commodity.

It is this international division of labor that explains Schumacher's "law of the disappearing middle." Appropriate, intermediate technology does not exist unless special efforts are made to create or preserve it, because in core countries the division of labor compels industry to employ ever more advanced technology. Commodities produced by such capital-intensive and ecologically violent technologies are competitive internationally. In the less developed peripheral economies similar but not quite as advanced technologies are imported and applied in urban manufacturing plants and in rural extractive industries. This is done so that the minerals and at least some of the manufactured products can be traded in the world market. The remaining 80 percent or more of the population living in the rural areas continue to use their traditional low productivity technologies. The periphery, then, has very simple and very advanced technology, but no middle. The core uses only the most recent technology, without a middle or a simple end of the spectrum.

Capitalist Commodity Production

> The market therefore represents only the surface of society and its significance relates to the momentary situation as it exists there and

then. There is no probing into the depths of things, into the natural or social facts that lie behind them. In a sense, the market is the institutionalization of individualism and non-responsibility.

E. F. Schumacher, *Small Is Beautiful*

The second basic structure of the MWS is that economic activity is organized along capitalist lines wherein privately owned firms produce commodities for private profit in the world market. Thus, the basic economic actors in the MWS are capitalist firms. As the writings of Frederick Lane have demonstrated, in the early formative centuries of the MWS when the reigning economic doctrine was mercantilism, there was not a clear-cut distinction between governments and firms. (Consider, for example, how Spain and Portugal ran their overseas empires.)[9] In time, and particularly after the 1776 publication of Adam Smith's *Wealth of Nations* and the beginnings of the so-called industrial revolution in England, economic doctrine and practice favored a sharp separation of the functions of government and business. The evolution of business organization has moved from the joint stock company to the privately held family business, to the modern corporation, and on to today's transnational corporations (TNCs) that dominate the world market. To be sure, state-owned corporations predominate in socialist countries and are found in many LDCs and in some core countries (for example, France's Renault auto firm). Nevertheless, capitalist TNCs undertake the vast majority of international trade and investment, thereby compelling state corporations to follow the same rules in their international transactions.

As capitalist enterprises, TNCs obey the "laws of capital," which have at their heart the endless search for profit so that capital may accumulate, or in other words, so that production can continually expand. Technology is vital to the capital accumulation process, and the technology favored is always capital-intensive. For any TNC or other capitalist firm, labor is nothing more than a factor of production like land, capital, and technology. The views of a leading British industrialist, chairman of the General Electric Company (U.K.), are typical: "People are like elastic, the more work you give them, the more they stretch."[10] Business practice and conventional economic theory both emphasize that profits and accumulation occur because of cost-effective competition. Since the nineteenth century, the expansion of democratic politics in core countries and the organization of workers' unions and political parties have meant that labor cannot be disciplined by forced unemployment; hence, the cost of labor has tended to increase regularly. The answer to such a situation is obvious—expand the corporation to achieve economies of scale and replace labor with machines via the introduction of ever more advanced capital-intensive technology.

As part of the expansion answer, modern corporations in the twentieth century have transformed themselves into transnational corporations. If in their overseas expansion TNCs adapted their production functions to local factor endowments and culture, the movement for intermediate and appropriate technology might never have come into existence. But for many reasons consistent with conventional economic theory and established business practice, TNCs tend not to adapt. Rather, they carry their capital-intensive, large-scale technology with them wherever they go. This may not be much of a problem when the move is from one core country to another, as when the General Motors Corporation sets up business in Germany. However, when a core-based TNC invests in a peripheral economy, as has Germany's Volkswagen in Brazil, the results are mixed at best. Brazil may record an increase in GNP and some industrial exports if the cars are sold elsewhere in Latin America. These are pluses. On the minus side, a capital-intensive technology has been introduced that inhibits the growth of a Brazilian automobile industry and provides a few highly paid jobs for urban Brazilian workers while it does nothing for the impoverished rural masses except, perhaps, to increase their dissatisfaction when they see Brazilian Volkswagens that they will never be able to buy.

Two further features of capitalist commodity production favor the replacement of men with machines. It must be recognized that one of the principal problems of production is that of coordination and control of the workplace. Machines do not bargain collectively, they do not strike, they do not talk back or think for themselves, they do not practice absenteeism or slowdowns. Second, money spent on wages is money gone, whereas money spent on machines enhances the net worth of the firm in the form of plant and equipment. From the point of view of capital, the ideal plant would have no workers at all, or very nearly none!

The dominance of the capitalist mode of commodity production, organized by TNCs, in the international political economy is clearly not neutral as regards appropriate technology. These enterprises are literally driven to revolutionize their techniques of production with a view to cost reduction and business expansion. They constantly develop ever more capital-intensive, large-scale technologies and transfer these technologies around the world, first among core countries and then from the core to the periphery. In a similar fashion, to compete internationally, state enterprises in so-called socialist countries and in newly industrialized countries such as Mexico and South Korea are driven to follow the technological innovations of the TNCs.

The State System

The third structural constant of the modern world system—its political organization into a system of over 160 sovereign states—is also not neutral as regards appropriate technology. Just as capitalism began to emerge out of European feudalism in the long sixteenth century, so did the modern nation-state. From the near political anarchy of the late Middle Ages in Europe there arose absolutist states that were in turn transformed into the nation-states we know today.[11] This process of state building was closely associated with international warfare. Success in war was based in part upon national wealth, hence the state-builders' interest in economic development and overseas expansion. Success in war was based on efficient political and military organization, hence the interest in bureaucratic rationalization. Success in war also was based on having superior armaments, hence state-builders' continual interest in the science and technology of weapons systems.[12]

The international state system concerns politics and power. It is an intensely competitive system characterized by struggles for dominance and survival. Survival is enhanced and the possibility of dominance arrives if a state can combine political and economic power sufficiently to thwart all possible rivals, individually and collectively. Such a state can then be called a "dominant power" or a "hegemonic state."[13] In the history of the MWS since 1450 only three hegemonic states have existed, although many others such as Spain under Philip II or France under Louis XIV have sought hegemony. These hegemonic states were the United Provinces (Dutch) from about 1600 to 1675, Great Britain from about 1763 to 1873, and the United States from about 1920 to 1970.[14] The political-economic histories of these three world powers illustrate at least three ways in which state building and the state system affect technological choices.

The first and most obvious impact is on military technology and its spinoffs in the private sector economy. Here again the impact has been to replace men with ever more sophisticated war machines (as discussed in the chapter by Bohlin; see pp. 109-19). This trend has reached the absurd level where the nuclear superpowers have the potential to extinguish the human race, if they ever do engage in an all-out thermonuclear war.[15] That so much state-sponsored technological innovation has the explicit intention of doing violence in war is related to the observed tendency of peaceful industrial technology to do violence to the environment and to workers' lives. During their periods of hegemony, the Netherlands, Britain, and the United States were both the strongest military powers and the most advanced capitalist economies. Their political and economic organization and their use of military technology were models for all other states seeking profits and power. Technological diffusion is therefore not neutral as regards the technologies transferred, nor is it apolitical.

It is less well recognized that states are active participants in the international economy. The capitalist world system combines a single world economy with a plurality of competitive states in a paradoxical structure.[16] Within the world market, firms seek profit and capital accumulation according to market forces such as comparative advantage and supply and demand. Firms prefer to let market forces operate as long as profits and growth result. However, if the operation of the market does not have this happy outcome, then firms often turn to the only actor that can control markets, the state, in order to secure their profits by political-military intervention within the world market. This pattern results in the basic dynamic of the MWS: "The functioning of a capitalist world-economy requires that groups pursue their economic interests within a single world-market while seeking to distort this market for their benefit by organizing to exert influence on states, some of which are more powerful than others but none of which controls the market entirely."[17] There are many ways in which states can intervene in markets, military force, imperialism, and colonialism being obvious examples. For the purposes of this chapter, the less obvious state subsidization of research and development of new technologies is central. All states now do this, with over 90 percent of research and development being done in a few industrial core states. Consider, for example, the vast amount of research and development done at public, state-supported American universities that ultimately is applied by American firms in industry and agribusiness.[18] The bias in state-supported research and development is usually for large-scale, capital-intensive technology whether the intended application is civilian or military. Control of such technology is a very real form of political and economic power, which is what states are all about. No wonder, then, that states do not support appropriate and intermediate technology very much. This is equally true of developed core states and of less developed peripheral states.

A final way that state building and the state system affect technology and technological choice is through the state enterprise. There is a clear trend among core and peripheral capitalist states to establish state-owned firms. Among the peripheral LDCs such policies give leadership to the state in the economic development of the country. In core states this results in the organization of capital on an efficient and large-scale basis so that the firms and the national economy will be competitive internationally. In both rich and poor countries there also are political and social objectives involved in the merger of state and capital in the guise of the state firm, but in my view these are not as essential as are the reasons mentioned above. This form of state interference in the operation of the market has been called neo-mercantilism, and not without reason. Whatever name is given to the state-capital merger, it is a growing trend because certain core states—Japan and France in particular—have demonstrated their success with this strategy for penetrating the world market. Most production for the world market is capital-intensive and large-scale, as we have repeatedly stressed. Hence, this role of the state in the MWS also is biased against intermediate and appropriate technologies.

Much more could be written about the structures and processes of the MWS, including the cyclical processes affecting control relations between the core and the periphery (colonialism, decolonization, neocolonialism), affecting the distribution of power among core states (hegemony or balance of power), and affecting production (Kondratiev's long wave). One also could examine trends in the system in more detail such as the accumulation of capital in ever larger firms, the increasing number of states in the system, and the system's periodic expansions to take in new populations and territories (e.g., the seabed and space). Such discussions would divert us from the main task of this chapter, to examine how the international political economy affects both the choice of technologies and the very existence of alternative, appropriate technologies. To do this, I shall now turn to the role of technology in the evolution of the MWS.

Technology in the History of the Modern World System

> In this way, modern technology became a class-bound phenomenon, the racing heart of corporate capitalism.
>
> David F. Noble, *America by Design: Technology, Science and the Rise of Corporate Capitalism*

Entire books have been written on this subject.[19] In just one section of one chapter the most that can be accomplished is to select several appropriate examples that demonstrate general tendencies. Since its origins in the sixteenth century the MWS has witnessed a long-term transformation involving ever more advanced technology. What is it about the MWS that stimulates continuous technological invention and innovation? This question can perhaps best be answered by examining two cases widely separated in time and space—the Dutch shipbuilding industry of the seventeenth century and the application of science to industry via the emergence of an engineering profession in the United States after 1850.[20]

The Dutch (the United Provinces) were the hegemonic power of the seventeenth century. It would appear that hegemony, with all its advantages for the pursuit of profits and power, arises in a sequence. Production efficiencies in agricultural and industrial activity make it possible to bring attractive, low-cost commodities onto the world market. Such productivity advantages lead to the growth of commercial leadership in world trade in these commodities. Commercial leadership can then be translated into financial dominance. Leadership in production, commerce, and finance produces super-profits that, when taxed or borrowed at low interest rates, permit the growth of a strong state apparatus, including the military. By means of the interdependence of politics and economics, strong states follow policies that aim to enhance commerce and finance and, to the extent that they support research and development, productivity as well. This was certainly the case in the Netherlands 350 years ago.

While Dutch agriculture was remarkably productive for the period, its industrial sector was even more advanced. The Dutch led the world in the production of the two major industrial products of the time—textiles and ships. In shipbuilding, the Dutch were most famous for their flyboats, or fluits, which provided cheap and reliable commercial transportation throughout European waters.[21] The flyboat was an outgrowth of several centuries of Dutch herring fishing in the North Sea and beyond. Thus, the gathering of a staple food was the basis for technological innovation in the shipbuilding industry of Holland and the other Dutch provinces.

By the early 1600s this industry was technologically very advanced, employing capital-intensive methods to construct merchant ships in a manner approaching mass production. The key elements in this technology were the use of standardized, repetitive methods of production and many labor-saving machines such as wind-powered sawmills, powered feeders for saws, blocks and tackles, and huge cranes to move the heavy oak timbers used in construction.[22] By midcentury the Dutch were building comparable ships 40 percent to 50 percent cheaper than their nearest competitors, the English. They had six cost advantages: the skill of Dutch shipwrights, economy in the use of materials, labor-saving devices, large-scale standardized production, large-scale purchasing of materials, and cheap transportation of construction materials from the Baltic to the Netherlands on Dutch ships. As Wallerstein comments, "Of these advantages, the first three may be seen as the technological advance of the Dutch, and the second three as the cumulative advantage of being ahead on the first three."[23]

But not only were Dutch ships cheaper to build, they were cheaper and more reliable to operate. Each of them needed a normal crew of 18 hands, whereas the ships built by competitors usually required from 26 to 30 crew members. This labor-saving design increased productivity and made it possible to have a healthier and better-fed crew than on the ships of other nations. In sum, Dutch ships were cleaner, cheaper, and safer.

The technological advantages of the Dutch shipbuilding industry had cumulative effects on Dutch commerce and finance. Dutch freight rates were by far the cheapest of the time. Indeed, Dutch merchants could sell Baltic goods in England more cheaply than could English merchants. As a consequence, Dutch shipping expanded greatly so that "as of 1670, the Dutch owned three times the tonnage of the English, and more than the tonnage of England, France, Portugal, Spain, and the Germanies combined."[24] Given that Dutch-built ships also were very common in their competitors' merchant marines, one can appreciate the dominance achieved by this remarkable people and their naval technology.

As mentioned, this technology of ship construction resulted in cheaper freight rates. Cheap rates led to Dutch control of trade, their own and that of other countries as well. Large volumes of trade produced economies of scale that reduced insurance costs which further cheapened rates, stimulating more trade and more ship construction. The cumulative advantages (i.e., positive, amplifying feedbacks) that derive from a technological breakthrough seldom have been more apparent.

Furthermore, not only did the Dutch export technologically advanced products like their flyboats and textiles, they exported technology directly. As the most technologically advanced society of its time—a fact recognized, admitted, and envied by the French and English—the United Provinces were a model for other socieities that were generally happy to receive Dutch investments, as in the Swedish iron industry, and to welcome Dutch engineers and artisans to drain their swamps, to upgrade their textiles, and to modernize their agriculture.

Although Dutch hegemony rested upon technological advances that occurred in the late Middle Ages in the Low Countries, the Dutch were able to apply and extend these advances only because of revolutionary social changes in the Netherlands during the Reformation and the eighty-year-long war of independence against the Spanish. If there is a message about the role of technology in the rise to dominance of the Dutch, it is that while technological advance may produce cumulative and long-lasting economic and political advantages, this will not happen unless technological innovation is combined with social change in such areas as property rights, labor recruitment and control, education, and government that favor the adoption and diffusion of the innovations. This happened in the United Provinces between 1560 and 1640 as it did in America between 1850 and 1950.

The "second industrial revolution" of the late nineteenth and early twentieth centuries in such industries as chemicals, electrical products, and automobiles differed from the first such revolution in a fundamental respect. It was based heavily upon the application of scientific knowledge to industry through the work of a new profession of technological experts, university-trained engineers. This linkage of science, technology, and large-scale capitalist industrial enterprise is the subject of David Noble's pioneering book, *America by Design*.

Engineering as a profession did not exist before the second half of the nineteenth century. Technological innovators before this time were normally mechanics and artisans, often with little or no formal education but with abundant practical experience. Such men often became capitalist entrepreneurs running their own enterprises and carrying most of their firms' operations in their heads. During the nineteenth century this form of the firm was gradually replaced by large, impersonal corporations producing ever more technologically sophisticated products. Noble shows that engineers were a vital aspect of this transformation of industry.

Central to this process was the transformation of the job of engineering into a profession similar to medicine with its requirements of advanced university education and rigorous entry standards. In the middle and late nineteenth century American engineers were divided into a "school culture" wing and a shop-floor wing emphasizing practical, on-the-job training. Scientific advances, their technological applications in industry, and the clear status and economic advantages associated with professional rank made it possible for the school culture wing to establish university education in engineering as a prerequisite for entry into the profession. Associated with and following this transformation, engineering schools were established throughout the nation —such as the Massachusetts Institute of Technology in 1862, Purdue University in 1865, and the California Institute of Technology in 1891.

These changes had implications for both universities and corporate capitalism. The universities acquired quasi-monopoly control of technical knowledge, replacing the shop-floor, rule-of-thumb methods of earlier generations of engineers. Universities and corporations were now linked by means of engineering education and the applied scientific research conducted at universities. Since many engineering schools and colleges were part of state-financed public universities, this shifted to the state part of the social costs of capitalist production —personnel training and much research and development. Universities in the United States, having established their leadership role in the development of advanced technology for industry, became a basic component of the capitalist mode of production within the larger MWS.

The application of science-based technological invention and innovation made possible a profound advance in the forces of production in nineteenth-and twentieth-century America. The new forces of production were organized by a new form of the firm, the modern national corporation. The rise of corporate capitalism made corporate planning possible and replaced the market anarchy of competitive, small-firm capitalism of an earlier epoch. Modern corporations were more stable and enduring. Their form of organization and scientific management procedures made possible the regulation of production, prices, and distribution.[25] The new engineering profession played a key role in this change.

As agents of modern technology and corporate capitalism, engineers did not restrict themselves to the technical side of the production process. With the rise of specializations such as industrial engineering, systems analysis, operations research, and scientific management in general, engineers increasingly left the shop floor and moved into corporate management positions. Initially, engineers contributed to the transformation of the forces of production through their research and applications of scientific knowledge. Now, as managers and as systems analysts, they designed and transformed the *social relations of production* to accommodate and facilitate the new technologies by structuring the labor force, planning work, and designing entire vertically integrated production systems. As Noble comments, "Forces of production and social relations, industry and business, engineering and the price system—collapsed together in the consciousness of corporate engineering, under the name of management."[26]

If by the mid-twentieth century it was no longer meaningful to distinguish between engineers and managers, the distinction between workers and others involved in the process of corporate capitalist production had been sharpened. The engineers' application of modern technology to industrial production in the United States after the Civil War extended the division of labor far beyond anything envisaged by Adam Smith for his famous pin factory. The work process was split up into literally hundreds of separate, repetitive operations. Few if any factory workers could any longer understand the whole production process and where their task fit into that whole. Thus, one can say that if the early capitalists expropriated workers' property, the modern engineer-manager as an agent of corporate capital expropriated workers' technical knowledge.

This case aptly illustrates how technical advance shapes the institutions of the modern world. Revolutionary advances in the forces of production arising from technological innovation and invention do not endure and spread unless there are corresponding fundamental changes in the social relations of production. The organization of the transnational corporation, the development of an engineering profession and engineering education linking universities and corporations, and the degradation of work by means of scientific management were such fundamental changes.

These patterns are not happenstance but due to the inner logic of the MWS itself. As a system organized along capitalist lines, the totality of the production process involves production, distribution, consumption, and reinvestment. That is, the capitalist, whether an individual or a TNC, starts the process with a given amount of money (M, which may be borrowed or owned outright). With M the entrepreneur buys and therefore owns plant and equipment, raw materials, and the labor power of workers. Via a production function these are combined to produce a product or commodity (C). This commodity must be transported or distributed to its potential markets where it is then sold and consumed. The objective for the capitalist is to realize more money (M') at the end than he began with. If M' is greater than M, then the successful capitalist can spend the difference for his personal consumption, or he can reinvest it to expand future production. This circuit of capital ($M - C - M'$) describes the capital accumulation process in a nutshell.

For many reasons this circuit is open to disruption. Therefore, capitalist economies, including the world economy, manifest cycles of prosperity and recession, boom and bust. A frequent cause of economic downturns is a realization crisis, wherein the capitalist cannot realize M' because overproduction has glutted markets and commodities remain unsold. This is a frequent form of crisis because of the capitalist's contradictory tendencies to maximize production and to minimize his wage bill, thereby creating insufficient effective demand for his products.

When confronted with a realization crisis, capitalism's historic answer has been to reorganize capital via mergers and other mechanisms during the recession and then to expand via the intensification and extension of production. Intensification of production involves such changes as capital investment in labor-saving machinery to increase the productivity of labor. Extension involves penetration into new areas of the world economy. It is obvious that both extension and intensification depend upon the application of science and technology to the production process as a totality. The merger of scientific engineering with corporate capitalism is an instance of intensification. Revolutions in the technology of communications and transportation are part of the extension process.

The entire production process involves distribution and consumption as well as direct production. Thus, in his drive to accumulate capital the capitalist must be concerned with what Marx called "the turnover time of capital"—production time plus the circulation time of capital. With a given production time, if commodities are to be sold in local and distant markets, then the only way to reduce turnover time and hence speed accumulation, is to reduce circulation time by means of more rapid transportation and communication. As Marx wrote: "While capital must on one side strive to tear down every spatial barrier to intercourse, i.e., to exchange, and conquer the whole earth for its market, it strives on the other side to annihilate this space with time. . . . The more developed the capital . . . the more does it strive simultaneously for an even greater extension of the market and for greater annihilation of space by time."[27] The logic here is straightforward. In order to expand production as continuously as possible and to overcome production crises, capital is driven to extend itself throughout the world economy. In doing this, capital has a strong interest in the most rapid possible communication and transportation consistent with reliability. As a consequence, pressure for constant technological invention and innovation in the areas of transport and communication characterizes the MWS. The Dutch flyboat of the seventeenth century discussed above and direct satellite broadcasting examined by Wigand are but examples of this general process of technological change.[28]

This section has tried to demonstrate the central role of technology and technological change in the organization and history of the modern world system. While sheer human inventiveness brings forth a certain amount of technological innovation, I have tried to show that the organization of our world structures technological change in certain directions favoring accumulation by large-scale enterprises and by states, which are but a form of violence-controlling enterprises.[29] The needs of capitalist corporations and nation-states are for labor-saving and space-annihilating technologies. What does all of this imply for the intermediate and appropriate technology movement?

Appropriate Technology in the Modern World System

> Small countries are more beautiful.
>
> E. F. Schumacher, *Small Is Beautiful*

> However, it is not for me to talk politics.
>
> E. F. Schumacher, *Small Is Beautiful*

For a political economist one of the most striking features of the literature on intermediate and appropriate technology is that it seldom discusses the political implications of these technologies. This should not be surprising, however, because the entire relationship between science and technology has not yet been approached in such a fashion. As Cohen comments, "How incomplete is the social sciences' treatment of the political economy of science may be judged from the lack of any fully elaborated classical, Keynesian or Marxist analysis."[30] Yet such analysis is very much needed because the choice of technologies is not politically neutral; some individuals and groups may benefit from each choice and some may not.

One need not view the world in zero-sum terms to appreciate that the quantity (Q) of a good produced by any production function (f) involving capital (K), labor (L), and technology (t) in the form $Q = f(K, L, t)$ implies both a distributional formula and a potential product mix. This economic statement is also a political function because it follows that "it is a political question as to whose factors of production are to be used and rewarded, whose tastes are to be met and satisfied, and whose factors and tastes are not."[31] While this point is rather obvious, it is usually ignored by conventional economists. What is not so evident is that production functions are also political statements because government resources are used to sustain or subsidize any function *and to preserve the function in the face of political opposition if such exists.*

The movement in support of intermediate and appropriate technology advocates production functions that favor the poor over the rich, the rural population over city dwellers, the unemployed over the employed, the small-scale producer over the large corporation, and local, self-reliant cultural values over the interdependent cosmopolitanism of the world economy.[32] The movement would agree with Gandhi's belief that "the poor of the world cannot be helped by mass production, only by production by the masses."[33] Applications of appropriate technology with these objectives in mind imply fundamental political reform and restructuring so that development meets basic human needs rather than simple quantitative growth, so that LDCs are collectively more self-reliant rather than more integrated into the MWS, so that income distribution within and between countries is made more equal, so that donor states and agencies reorient their foreign aid policies in support of appropriate technology, so that new international agencies such as an IMAT (International Mechanism for Appropriate Technology) are created, and, one may infer, so that a new international economic order (NIEO) comes into being. All of this may be utopian. It is unquestionably highly political and likely to provoke sharp opposition if it ever appears to be happening.

It would seem that discussions of the nature and implications of appropriate and intermediate technology for too long have been dominated by economists and engineers. One searches in vain for writings that directly confront the international political-economic implications of the movement.[34] But how politically strong is the movement anyway? Not very, it would seem. The pathbreaking organization in the field, the Intermediate Technology Development Group Ltd., was founded only in 1966 as a nonprofit corporation and registered *charity* by means of Schumacher's donation of the honorarium he received for a newspaper article.[35] This organization's annual budget in 1975 was only eighty-thousand pounds sterling, and the total expenditure of all organizations involved in developing and diffusing appropriate technologies on a worldwide basis in 1975 was estimated to be a mere $10 million. Only one-half this amount, or some $5 million, was spent on appropriate technology research and development in 1975. In contrast, research and development for advanced, capital-saving technologies in that year was as much as $60 billion.[36] What can such a poorly financed and politically fragmented movement do to change a world in which 98 percent of all scientific research is undertaken in rich countries to address their problems, with little to no regard for the problems of the LDCs?[37] One is tempted to say,

"Not very much," but that would be defeatist. Rather, the appropriate technology movement must recognize that it *needs an appropriate political strategy if it is ever going to accomplish its objectives.*

An essential element in such a strategy would be to recognize where the movement for appropriate technology fits into the modern world system. The MWS is organized in the interests of the large, rich, and powerful actors in the system—states, transnational corporations, and socioeconomic elites.[38] While not directly confrontational, to the extent that the intermediate technology movement advocates production functions favoring the weak and poor of the world, the movement is anti-systemic in its thrust. That is to say, for intermediate and appropriate technologies to be used extensively to do the world's work it may be necessary to restructure the world in such a manner that states and TNCs cease to be the basic political-economic actors in the system. When stated in such bold fashion, the enormity of the task at hand becomes staggering.

Thus, besides recognizing the inherently political aspects of any choice of technology and the stacking of the deck within the MWS against small, simple, capital-saving, nonviolent technologies, it is also necessary to recognize that attitudes toward appropriate technologies "will differ depending on whose technology it is, on which specific technological advance we evaluate, on which portion of humankind is speaking or is represented, which class, which race, which tribe, which generation, which sex, and at which cultural place the evaluator stands."[39] The discipline of political economy suggests that the perceived impact of a proposed technological choice on the evaluator's profit- and power-seeking activities will interact with these variables to determine the evaluator's ultimate attitude and technological preferences.

Four attitudes toward intermediate or appropriate technology can be identified: (*a*) rejection of the concept, (*b*) acceptance of the idea in principle, (*c*) active involvement in knowledge mobilization and experimentation, and (*d*) willingness to apply the concept as a normal part of business, administrative, and community activity.[40] Schumacher and his followers, being missionaries, believed that these attitudes formed a sequence of gradual acceptance of their message, with the LDCs now moving from (*b*) to (*c*), but with the rich countries somehow still stuck at (*a*).[41] They have not recognized sufficiently that for state and corporate managers an attitude and policy of rejection or antagonism is both rational and self-interested, at least in the short run. The movement has not yet come to terms with the seemingly obvious fact that control of advanced, large-scale, labor-saving, and space-annihilating technology places enormous power—political and economic—in the hands of technological elites and the institutions they direct.[42] Why should such elites ever change their hostile attitudes to appropriate technologies? To argue that it is in their long-term best interest is both futile and ingenuous.

It would appear, then, that the movement ought to focus more of its efforts on the soft technology side of things, on the development of a political economy of appropriate technology that is self-consciously materialist in orientation, and on studies of social change, particularly as it relates to acceptance and diffusion of appropriate and intermediate technologies. Also needed are direct political actions in alliance with other anti-systemic forces and movements so that at least some of the world's more powerful institutions can be captured or controlled by supporters of the movement. Little effort, however, should be spent on trying to convert or capture TNCs. It is the governments of Third World states and international organizations linked to the United Nations that offer more promising prospects. Organizations dedicated to appropriate technology should openly support the movement for a new international economic order and forge coalitions and alliances with other NIEO supporters. Legitimating a basic human-needs approach to development and the idea that science and technology are the common heritage of humankind is as important as any engineering study of a particular intermediate technology.[43]

In seeking acceptance of these technologies by Third World governments the movement should not assume—as most development economists have—that these states are ruled by a unanimous elite coalition setting national policy, that they are politically quiescent, that organized labor is fully controlled, that the peasants are compliant and unable to be mobilized, and that the regimes are able and willing to use considerable coercion to achieve their objectives.[44] Monolithic states of this sort do not exist in the core of the MWS much less in its periphery. State-managers in both regions of the world system cannot be expected to embrace appropriate technology with open arms. Yet if elites in the Third World can see such technologies as linked to national objectives of self-reliance and development, and if linkages can be made to the rural masses and their authentic leaders, then appropriate and intermediate technologies might well become more widespread and ordinary.

The appropriate technology movement needs to transcend its liberal, voluntaristic mode of operation via alliances with other anti-systemic forces. It must become a political force within the MWS if technology with a human face is to have a future. It would be foolish to underestimate the strength and rational self-interest of the movement's opponents, that is, all of the elites and institutions favoring advanced, large-scale, capital-saving technologies. But the effort, though great, can be expected to be worthwhile. Schumacher was certainly right when he wrote that "the future cannot be forecast, but it can be explored."[45] If it is true that technological advance accounts for from 50 percent to 75 percent of the world's economic growth since 1770,[46] then it is reasonable to explore the possibility that the future development of the world's poor regions will be predicated upon technological advances appropriate to their needs. To ignore this promise would mean accepting a future in which two-thirds or more of humankind will continue to live in poverty, ignorance, and disease. Because appropriate technology suggests that the future need not be like the past, we have a choice to make. One can hope that the choice will be for humanity, and that this choice will be informed by an awareness of the political and economic realities that must be overcome.

4. The Transfer of Technology to Third World Countries: Political Problems and International Ramifications

Werner J. Feld

Perhaps nothing has disturbed Third World leaders more than the lack of technology and scientific know-how in their countries. They feel that the possession of such knowledge has invested the superpowers and the rich countries with enormous clout in the economic and political affairs of the world. They perceive technology to be closely tied to political and economic independence as well as to the provision of jobs for their people and economic development in general. Thus, it is not suprising that they want to change the rules of the game regarding technological and industrial property rights in accordance with the principles of the New International Economic Order (NIEO) and the Charter of Economic Rights and Duties of States. The United Nation's Conference on Trade and Development (UNCTAD) has been and continues to be the logical forum for these efforts.

The largest portion of the technologies obtained by less developed countries (LDCs) was transferred through the medium of transnational corporations (TNCs). These corporations either already had subsidiaries in the LDCs or established them, in part at least, for the transfer of particular technologies. For this purpose they also may have engaged in joint ventures with local enterprises or licensed the use of patents to indigenous firms. For these reasons it is evident that the issue of lacking but needed technologies and the political and legal problems associated with this issue are intimately linked with the research and development and other activities of TNCs in the Third World. This chapter will examine some of these problems, assess the progress made so far in their solution, and make some tentative projections for the future. While the primary focus will be on the role of the TNCs, relevant peripheral aspects such as the nature of industrial property rights as they affect Third World aspirations also will be discussed briefly.

Before embarking on our inquiry, it is important to keep in mind that the burden for the transfer of technology to LDCs falls on both the supplier and the recipient. As Harvey Wallender puts it aptly: "Technology transfer for development is like a game of catch. Much can be done to improve the ability of the pitcher and his toss. However, without a skilled catcher, the game will be incomplete or at best extremely difficult."[1]

Background

To improve the level of technological capability in the Third World, efforts in UNCTAD have centered on developing a code of conduct on the transfer of technology acceptable to both LDCs and the developed countries. Draft outlines for such a code were formulated early in 1975 by Brazil for the Group of 77 and by Japan for the Western industrialized countries. These outlines were discussed at the UNCTAD Trade and Development Board meeting in May of that year and were then expanded and revised in November 1975.[2]

The framework of the UNCTAD efforts can be seen best by quoting from Item 12 of the Provisional Agenda for the Fourth Session of UNCTAD in Nairobi in May 1976:

> A new phase is beginning in the developing countries—a phase marked by a radical shift of vision and the search for new policies. The peripheral policies of the past, involving minor modifications to existing forms of relationships, are being replaced by a search for fresh patterns drawing upon economic, social, and cultural resources indigenous to the territories of the Third World. The strengthening of national technological capabilities is assuming a central place in development plans and policies. Attempts are therefore being made progressively to loosen those ties with developed countries which hamper the attainment of this objective, and to move towards greater cooperation among countries themselves.[3]

UNCTAD's concern with the need for altering the existing international legal environment and practices in the area of technology transfer to meet the interests and perceived needs of the Third World dates back to the first UNCTAD Conference in 1964. However, despite some major studies undertaken by the UNCTAD Secretariat and the formulation of action programs for the governments of LDCs (including regional and interregional cooperation as well as action by advanced countries), little was done between 1972 and 1975 to implement these plans. In May 1975 an Intergovernmental Group of Experts on the Code of Conduct on Transfer of Technology, which was convened by the secretary of UNCTAD, reconsidered earlier plans on the subject. To push this endeavor along, resolution 3362, adopted at the Seventh Special Session of the General Assembly of the United Nations, stated: "All States should cooperate in evolving an international code of conduct for the transfer of technology, corresponding, in particular, to the special needs of the developing countries." The resolution set as a target date the end of 1977 for a code for the transfer of technology. Although this deadline has not been met, considerable progress has been made, as will be shown.

Meanwhile, as a collateral endeavor to the code on technology transfer, UNCTAD also has embarked on efforts to restructure the existing industrial property systems and specifically to revise the Paris Convention for the Protection of Industrial Property of 1883 as amended. These efforts were guided by a number of considerations, of which the following seem to be most significant:

(1) The importation of the patented product is not, as a general rule, a substitute for the working of the patent in the developing country granting it.

(2) More adequate provisions are required to avoid abuses of patent rights and to increase the probability of patents being worked in the developing country granting them.

(3) The introduction of forms of protection of inventions other than traditional patents (e.g., inventor's certificates, industrial development patents, and technology transfer patents) should be examined.

(4) The need for technical assistance to developing countries in the field of industrial property, and in particular for expanded access to and utilization of patent documentation by developing countries must be recognized, in order to facilitate the transfer, absorption, adaptation, and creation of suitable technology.

(5) An in-depth review of the provisions on trademarks should be carried out.

(6) There should be new and imaginative studies of possibilities of giving preferential treatment to all developing countries . . .[4]

To make the efforts for a code on technology transfer and for a review of the patent laws meaningful for most of the Third World, UNCTAD also has recognized the need for requisite institutional machinery and trained technical personnel. Without a proper infrastructure, the UNCTAD plans may conjure up a pleasant vista of the future, but the realization of the goal of rapid industrialization aided by new patent and technology transfer systems would be only a very dim prospect.

Third World Demands

In the first draft of a code for the transfer of technology, which was prepared for the Third World by Brazil in 1975, the preamble regards technology as "a part of universal human heritage to which all countries have the 'right' of access in order to improve the living standards of their people. All countries have therefore the 'duty' to promote the transfer of technology, whether proprietary or otherwise, on favorable terms" in accordance with the national policies, plans, and priorities of the developing countries. Indeed, an adequate transfer should become an "effective instrument for the elimination of economic inequality among countries and for the establishment of a new and more just international economic order." For this reason the code should be "universally applicable" and legally binding internationally.[5]

Of course, preambles tend to be hyperbolic, and in the body of the proposed code the tone is less strident. Emphasis is placed on the obvious right of home and host countries to regulate the transfer of technology through national legislation. Such legislation must assure protection of domestic recipient enterprises and prevent, in general, the displacement of national enterprises by foreign collaboration arrangements. Payments for technology are to be treated as profit whenever such payments are made to parent companies or other subsidiaries of a TNC or when the supplier and recipient companies "form an economic unit or have community of interests."

A long chapter of the proposed draft outline deals with restrictive business practices in the transfer of technology. About forty specific restrictions are enumerated that are seen as possibly having an adverse effect on the technology recipient. Some of these restrictions, such as the prohibition of horizontal cartel activities, were reasonable and might be supportable by TNCs and their home country governments; others were rejected by TNCs as unduly limiting the freedom of contractual agreements between supplier firms and recipient enterprises.

Another chapter of the draft requires numerous guarantees from enterprises supplying technology. These include adequate training of nationals in the use of the technology received and fair pricing for needed materials imported and for goods produced when they can be sold only to the technology supplier or any other enterprises designated by him. Priorities must be given to the employment of local research and development skills and experience and to the full use of technology already available in the recipient country. In addition, special preferential treatment is stipulated for enterprises in developing countries. Assistance also is to be given by the governments of developed supplier countries for the establishment of national, regional, and international institutions that can help the Third World nations in their quest for greater technological capabilities.

For the settlement of disputes arising from technology transfer agreements, the laws of the technology-receiving country are to apply. Only if these laws specifically permit recourse to arbitration in this field can the parties concerned submit such disputes to arbitration in accordance with procedures to which they have agreed.

Western Responses and U.S. Position Changes

Not surprisingly, a wide conceptual gap separated the Third World draft outline from the initial proposal of the industrially advanced countries, which insisted on *voluntary* compliance by TNCs and governments and on maximum freedom and the sanctity of contractual agreements. The bias of this proposal was clearly in favor of TNCs and the "proper" investment climate, although the basic needs of Third World countries for appropriate technology were acknowledged and the legitimacy of some of their demands was recognized. The problem of restrictive business practices was dealt with in one line, and the responsibilities of technology-supplying enterprises and their governments were couched in qualifying terms such as "to the extent practicable," "feasible," "appropriate," and "reasonable."

However, the revised draft outline on technology transfer submitted by the Western industrialized countries in November 1975 was much more comprehensive and positive, thereby appreciably narrowing the gap between the Third World and the Western industrialized camp. Perhaps reflecting changed United States foreign policy that manifested itself during the Seventh Special Session of the United Nations General Assembly in September 1975 and the apparent, though qualified, responsiveness of the Third World leadership to American policy suggestions, the revised proposal clearly stated that the development of indigenous technological capabilities in the Third World should be promoted, and that restrictive business practices adversely affecting the transfer of technology should be avoided. Moreover, details for the implementation of these objectives were stipulated by assigning specific responsibilities to technology source and recipient enterprises and governments.[6]

Of course, the revised proposal of the Western group, supported by the United States, continued to emphasize the confidentiality and proprietary nature of trade secrets and know-how acquired in connection with the transfer of technology. It opposed the view that the law of the recipient state should determine which legal rules are to be applied for the settlement of disputes arising from the transfer of technology and wanted the parties to pertinent agreements to choose freely the applicable law including arbitration. Nevertheless, some changes in the U.S. position began to show up in 1975 when Secretary of State Henry Kissinger addressed the Seventh Special Session of the UN General Assembly. He stated:

> The United States is prepared to meet the proper concerns of governments in whose territories transnational enterprises operate. We affirm that enterprises must act in full accordance with the sovereignty of host governments and take full account of their public policy. Countries are entitled to regulate the operations of transnational enterprises within their borders, but countries wishing the benefits of these enterprises should foster the conditions that attract and maintain their productive operation. The United States therefore believes that the time has come for the international community to articulate standards of conduct for both enterprises and governments.[7]

The new attitude of the United States was further explained in another speech by Kissinger, given at the opening session of UNCTAD IV in Nairobi. In this important address Kissinger outlined four actions the United States was prepared to take to remedy the technology gap. First, to promote the adaptation of technology to meet LDC needs, the United States would support a global network of research institutions devoted to this task. Second, to promote the worldwide diffusion of technological information, the United States would inventory its resources in this area and facilitate LDC access to repositories of such information like the National Agricultural Library. Third, the United States would expand educational and training efforts to provide a new generation of technological experts. Fourth and finally, the United States would support international efforts to make the transfer of technology more "effective and equitable."[8] Secretary of State Cyrus Vance continued this policy trend. In a major speech in March 1979 he said:

> the ability of people and institutions in the developing countries to obtain, develop, adapt, and apply technology is critical to most development problems. . . . We have facilitated access to the technology that is in the public domain and we have helped developing countries draw upon our advanced technologies—using satellites, for example, to develop their natural resources and improve their internal communications.[9]

Areas of Consensus

Perhaps aided by this more favorable stance toward Third World demands adopted by the U.S. government regarding technology proprietary rights and the transfer of technology, the labors of the Working Group of Experts on these subjects have made progress toward an agreed text of a code. The code consists of preamble and ten chapters. The latter cover definitions and scope of application; objectives and principles; national regulation of transfer of technology transactions; restrictive practices involving the transfer of technology and exclusion of political discrimination; guarantees, responsibilities, and obligations of parties; special treatment for developing countries; international collaboration; international institutional machinery; and applicable law and settlement of disputes. The scope of the code is broad, and its coverage includes patented and unpatented technology, turnkey agreements, and leasing of machinery. Transfer of technology is understood for purposes of the draft code as the transfer of systematic knowledge for the manufacture of a product, for the application of a process, or for the rendering of a service. The code applies only to *international* transfer of technology but covers a variety of enterprises going beyond TNCs and is addressed to all governments, irrespective of the economic and political systems of their countries. Indeed, parties to a technology transfer may be state-owned TNCs such as Renault. Another point needs to be emphasized: it has been agreed in principle that the technology transfer code will apply to intracorporate transactions, provided that the transfer passed national boundaries; however, details on this matter remain to be worked out.[10]

It is significant that major agreement has been reached on the objectives of the technology transfer code. They include:

> 1. To encourage transfer of technology transactions, particularly those involving developing countries, under conditions where bargaining positions of the parties to the transactions are balanced in such a way as to avoid

abuses of a stronger position and thereby to achieve mutually satisfactory agreements.

2. To facilitate and increase the international flow of technological information, particularly on the availability of alternative technologies as a prerequisite for the assessment, selection, adaptation, development, and use of technologies in all countries, particularly in developing countries.

3. To facilitate and increase the international flow of proprietary and nonproprietary technology for strengthening the growth of the scientific and technological capabilities of all countries, in particular developing countries, so as to increase their participation in world production and trade.[11]

Among the principles underlying the technology transfer agreed upon are the universal applicability of the code and cooperation of all states in such transfers to promote economic growth throughout the world, especially that of the LDCs, but irrespective of any differences in political, economic, and social systems. Technology-supplying parties, when operating in a country acquiring such technology, should respect the sovereignty and the laws of that country, show proper regard for the country's declared development policies, and endeavor to contribute substantially to the development of that country.[12]

Throughout the draft code, emphasis is placed on recognition of the sovereignty and independence of states and their sovereign equality. Hence, there is general agreement on the section dealing with national regulations regarding technology transfer transactions, although some changes in the provisions could well be made before this code is accepted by all parties.[13] In particular, concerns have been voiced in Western developed countries, and especially the United States, as to whether legal standards of LDCs embodied in their national laws and regulations will conform to those normally prevailing in the Western World.

The most controversial principle of the code on the transfer of technology, as it is also with respect to the general code of conduct for TNCs, remains its legal nature. All LDCs have advocated persistently that the code should be legally binding in every respect and should cover transactions between all affiliated enterprises, but that responsibilities addressed to enterprises need not be balanced by responsibilities of governments. A follow-up mechanism was to be created to supervise the implementation of the code.

The U.S. position, on the other hand, has been consistent in its claim that the code must be voluntary in nature and appropriately balanced in reference to the responsibilities of enterprises and governments. In addition, national provisions addressed to enterprises by LDCs regarding technology transfers should be based on standards and practices that do not contravene the principles of international law.[14]

Initially, members of the State Department Advisory Committee on International Investment, Technology, and Development appeared to be opposed to any meaningful follow-up mechanism to the implementation of a voluntary code. However, in 1979 the U.S. position seemed to soften and move toward the acceptance of follow-up machinery to review the working of the code. Other developed countries have expressed similar views and, during a session of the UN Conference on an International Code of Conduct on the Transfer of Technology held in Geneva, October 22–November 2, 1979, the following compromise emerged between the U.S. and LDC positions. The code would be voluntary, but after a period of time (four or six years) its legal status would be reviewed through some kind of follow-up machinery.[15] However, as we will see, only the broad concepts of that machinery have been agreed upon. Consensus on the details of this machinery is likely to be a very difficult task since, in discussions in the State Department Advisory Committee working groups, opposition was expressed to any effective follow-up machinery, even on the OECD Guidelines for Multinational Corporations, because it would undermine their voluntary character.[16] Another troublesome issue may be agreement on which agency would be in charge of the follow-up process.

Finally, concern was expressed that some provisions of the code would find their way into national legislation and could then become a part of general international law that courts may apply for settling disputes.[17] Of course, the enactment of national legislation on the international transfer of technology can be carried out now and is specifically recognized by the draft code that sets forth a number of desirable objectives and principles. These include the promotion of a favorable and beneficial climate for the international transfer of technology, the encouragement of mutually agreed upon, fair, and reasonable terms and conditions, and other factors generally congenial with U.S. interests.[18] Perhaps two of the most significant clauses agreed upon in support of the U.S. position state that each country "should ensure an effective protection of industrial property rights granted under its national law and other related rights recognized by its national law," and national measures should "be consistent with . . . international obligations."[19]

Prohibition of Restrictive Business Practices and Other Obligations

Considerable progress has been made to close the gap on the troublesome issue of restrictive business practices (RBPs) involved in the transfer of technology. The LDCs have lowered their list of such practices from forty to twenty, and the United States and other Western developed countries now recognize a list of sixteen of these practices. In fact, some of the provisions now have an agreed text, and in quite a few others the wording is relatively close, a major difference being the inclusion of qualifying words such as "unreasonable" in the Western proposal. Some progress on the substance also has been made on the obligations of the parties to technology transfer.[20]

Major agreement has been reached on fourteen practices. They include grant-back provisions by the technology-acquiring party to the supplying party; restraints on the acquiring party from challenging the validity of patents and other types of protection involved in the technology transfer; restrictions on the freedom of the acquiring party to enter into sales ("exclusive dealing"); restrictions on research and the use of personnel by the acquiring party; restrictions on the prices to be charged by the acquiring party; restrictions on adaptations or innovations by the acquiring party; restrictions on adaptations or innovations by the acquiring party that require acceptance of additional technology not wanted by the acquiring party ("tying arrangements"); export restrictions on shipment of goods by the acquiring party; restrictions or payments after the expiration of acquired industrial property rights; and other limitations. Several other, highly complex issues involving RPBs will need to be negotiated. They include the exhaustive or unexhaustive nature of the list of practices in this chapter of the code and the treatment of technology between affiliated enterprises. Agreement also is lacking on additional RBSs.[21]

The chapter dealing with Guarantees/Responsibilities/Obligations covers both the precontractual phase, i.e., the period when potential parties are negotiating the terms of the transactions, and the contractual phase after the parties have entered into an agreement on the transfer of a particular technology. Among the obligations specified in the draft code is the responsibility to clearly relate the acquisition and delivery of technology to the governmental economic and social development objectives and the observation of fair and honest business practices. However, there is as yet little consensus on how far respect should go for confidentiality of trade secrets, secret know-how, and other confidential information that the technology-acquiring party might receive from the supplying party. Other aspects of the transfer accord also are lacking a consensus, but a beginning has been made in reaching agreement on this difficult topic.[22]

Miscellaneous Sections of the Draft Code

The sections of the draft code on special treatment for developing countries and on international collaboration posed few problems. The provisions are mainly programmatic, and since on both subjects the objectives to be attained are congenial to Third World and developed countries, rapid progress was made in formulating the pertinent clauses.[23]

A tentative, rather comprehensive text on international institutional machinery has been elaborated to follow up the implementation of the code, but agreement on the precise wording on many of the pertinent clauses has not yet been achieved. As in the case of the general code of conduct for TNCs, the follow-up machinery is a sensitive and somewhat controversial issue. Agreement exists that it should be an intergovernmental body, for which the UNCTAD Secretariat should provide the needed secretarial services. Furthermore, after a number of years (four or six) following the adoption of the code, a review of all aspects of the code is to begin. The character and mandate of the review conference is still in dispute. It is noteworthy that this body or any subsidiary organ is prohibited from acting as a tribunal or otherwise passing judgment on the activities or conduct of individual parties or governments in connection with a specific transfer of technology.[24]

Finally, concerning "applicable law and settlement of disputes," agreement so far has been elusive. Indeed, it has not even been possible to prepare a composite text as a basis for further negotiations, since the approaches of the regional groups have remained divergent.

An effort collateral to the elaboration of the code has been the call for and finally the convening of a UN Conference on Science and Technology for Development in August 1979 in Vienna. However, despite much rhetoric and expenditure of money for still another conference, not much has been accomplished. Nevertheless, an interim fund of $250 million was created to be spent over a period of two years for the improvement of science and technology for development. Beyond that, the Vienna conference made little progress in overcoming the still existing gaps between the positions of the Western industrialized nations and the developing countries on technology transfer. Despite some rhetoric to the contrary, the LDC representatives expressed their disappointment over the outcome.[25]

The Industrial Property System

As already noted, the demands of the LDCs for basic changes in the patent system and, more generally, the protection of intellectual property so far have not been successful. The World Intellectual Property Organization (WIPO) formed a Group of Experts on the Revision of the Paris Convention that met several times in 1975 and 1976. This group since has been renamed the Preparatory Intergovernmental Committee (PIC) and organized meetings in 1978 and 1979. A Conference of Revision was convened in February 1980.[26]

The U.S. position is that the Paris Convention by its nature and aims cannot be revised to solve all the problems of LDCs with respect to the transfer of technology. The basic principles of the Convention must not be tampered with. However, the United States is not at all opposed to the modification of the Convention with a view to providing special benefits to Third World countries, while, at the same time, protecting the interests of the developed countries.[27]

In the 1978 and 1979 PIC meetings tentative agreement was reached on (1) supplying patent information from one national patent office to another upon request; (2) endeavoring to use industrial property whenever possible to enhance the development process; and (3) authorizing the Assembly of the Paris Convention to recommend industrial property projects for the WIPO program for legal-technical assistance to LDCs. But significant areas of disagreement remain. For example, the concept of preferential treatment without reciprocity for nations of LDCs with regard to patent fees and/or priority periods has not been accepted by the United States and other developed countries. And no agreement has been reached on the procedural question as to whether the Convention should be revised by unanimity, as in the past, or by some qualified majority.

Prospects for Outstanding Issues

The Legal Character of the Code

The issue of the legal character of the prospective code permeates almost all provisions of the document. The UNCTAD Secretariat has explored various solutions that might form the basis for compromises. For example, it has considered whether the code might contain mandatory and nonmandatory provisions in all of its chapters. Another possibility would be the adoption as a set of voluntary guidelines through a UN General Assembly Resolution. Subsequently, some chapters or parts of the code could be adopted as a treaty by states wishing to do so. Finally, the parties to particular transfer of technology transactions may agree to be legally bound by the code.[28]

I believe that at present none of these possibilities will appeal to the Western industrialized countries because in one way or another they suggest that the code will become a legally binding instrument in an incremental manner. Perhaps the best way out of this dilemma may be adoption of the code as voluntary guidelines by a resolution of the General Assembly with a review conference after a fixed time to further explore the legal nature issue. Whether this will be accepted by the LDCs (Group of 77) is, of course, questionable, but it would be a step forward.

The "International Transfer" and "Parent-Subsidiary" Issues

Although, as we have seen, there is agreement that the transfer of technology transaction falling under the provision of the code must be "international," meaning that the technology is transferred "across national boundaries," several problems persist. Many states apply the principles of the code of technology transfer transactions taking place within their national boundaries by means of enacting appropriate legislation. The Western countries seem to be in favor of such an extension. The Group of 77 and the Communist countries are opposed. However, there appears to be agreement that the "international" element can be satisfied if the parties to a transaction are located in different countries, even though transactions take place within one country.

How far does the code apply to transactions between the parent company and the subsidiaries of a TNC? There is agreement that the provisions regarding arrangements for technology transfer and restrictive business practices apply basically to dealings between independent entities. For the Group of 77 these provisions also are to govern all intraenterprise transactions, but the Western countries argue for only a qualified application depending on individual circumstances. To resolve these problems, relevant compromises have been proposed by the president of the Conference on an International Code of Conduct in the Transfer of Technology. These compromises appear to have a good chance of acceptance by all parties, although additional bargaining may well be necessary.

Conceptual Problems on Restrictive Business Practices (RBPs)

Although much progress has been made to overcome the initial gap about RBPs between the LDCs and industrially advanced countries, some fundamental conceptual differences persist. The Group of 77 wants to eliminate all practices that, whether restraining competition or not, are perceived as prejudicial to the economic and social development of the countries concerned, and, of course, in particular that of the LDCs. Most developed countries, on the other hand, view the dismantlement of RBPs as grounded in their restraints to trade and to the international flow of technology, and therefore do not want to accept those practices that fall outside of this category. Again, a likely compromise has been proposed by the conference president. This proposal assumes that the provisions on restrictive practices in the code are sufficiently explicit regarding their scope and operation to achieve the objectives of the code in a satisfactory manner without a conceptual interpretation. However, the basic criteria in determining an exclusion of particular practices should be whether, in an individual case, the overall effect of a technology transfer transaction, on balance, is in the national interest of the technology-acquiring country.[30] The chances for compromise on these issues seem to be good.

Applicable Law and Settlement of Disputes

Disagreement reigns in the area of what law should be applied in the case of disputes, and the outlook is dim for finding a compromise. The group of Western countries takes the position that the parties to a transfer of technology transaction should have the freedom to choose the applicable national law and the national forum before which disputes will be brought. This group also recommends the settlement of disputes of arbitration and the recognition and enforcement of arbitral awards under the UN Convention for the Recognition and Enforcement of Foreign Arbitral Awards. The Group of 77 asserts that the law of the acquiring country is the law applicable to matters relating to public policy and that any clause to the contrary is void. Recourse to arbitration is permitted only if the acquiring country does *not* have explicit rules to the contrary.

Various attempts by conference chairmen have been made to resolve the differences between the LDCs and the Western advanced countries, but so far to no avail. The obvious reason is that this issue is not only very complex, but also politically sensitive with emotional nationalistic overtones. Hence, it has been suggested that it may be preferable to postpone discussions on this problem until the review of the code by a later conference, and that, in the meantime, the international institutional machinery should concern itself with preparing an acceptable set of provisions on this subject.[31] But whether this strategy will prove to be successful remains to be seen.

Assumptions and Interests

The technological efforts of the Group of 77 in UNCTAD have been based on three fundamental assumptions: (1) advanced technology and scientific know-how in the industrialized countries of the "North" have produced a high level of economic development; (2) acquisition of technology and know-how by the LDCs will materially aid their economic development and will work toward eliminating inequality among countries; (3) technology is part of the universal human heritage, and all countries have the right to its access.

This set of assumptions suffers from a number of flaws. Obviously, assumption three is strongly contested by the Western countries. Moreover, the mere adaptation of existing technology to perceived needs of LDCs would entail considerable costs that include development of national skills in the Third World. Assumption two ignores the need for careful differentiation of the kinds of technology that should be imported; it also fails to take sufficient account of the disparity in economic levels and interests of LDCs. Although throughout the UNCTAD negotiations the LDCs insisted on *appropriate* technology, the selection of such technology might be extremely difficult considering cost and indigenous factors and requires the highest technological expertise on the part of the authorities of the acquiring country. While the UN is prepared to offer vast quantities of technical advice and to send experts to those LDCs seeking to acquire technology, the training of such experts is costly and time-consuming. As Rachel McCulloch has pointed out in a draft paper, since past research and development are a sunk cost and existing technology can be used in the Third World at little *further* costs in many instances, there is a tradeoff between incurring *further* costs to tailor products and processes to Third World markets and production conditions versus incurring higher social and economic costs by applying the inappropriate technology in essentially unchanged forms. Once it is recognized that appropriateness has its own price, it becomes clear that in some cases appropriate technology is an inappropriate choice—"it may sometimes make more sense to wear the hand-me-down unit as it is than to pay a tailor to achieve an elegant fit." This is certainly a fine prescription, but it overlooks the political sensitivity and emotional nature of the whole issue.

Another factor tending to promote the acquisition of inappropriate technology is national legislation and policy of LDCs based on inadequate information. Hence, such policies may be too restrictive, and the inadequacy of information may lead to choices that are not necessarily in the best interests of the more backward countries. Examples of the so-called NICs (newly industrialized countries) are Mexico, Brazil, Taiwan, and South Korea. It is interesting to note that the disparity of interests has not disrupted the basic solidarity of the

Group of 77 in the UNCTAD negotiations, although under the surface some tough bargaining may go on when common negotiating positions of the group are hammered out in its dealings with the Western group of countries. Indeed, at times the UNCTAD Secretariat has become a party itself in the negotiations with its own interests, perhaps partly motivated by bureaucratic, tactical considerations, although these interests normally are fairly close to the objectives pursued by the LDCs.

Success or Failure in Negotiations

Success or failure of the negotiations regarding a code on the transfer of technology depends on a number of factors. Although, as we noted, the gap between the Group of 77 and the Western industrialized countries has been narrowed, much will depend on the future foreign policy positions of the United States since U.S. TNCs normally supply the major part of technologies to the LDCs. The positions of Third World leaders also are very important; some of these leaders seem to have shown more down-to-earth attitudes, contributing to enhanced prospects for overcoming some of the divergences in positions. It may well have been with this progress in mind that the group of Western industrialized countries issued a statement at the conclusion of the Nairobi UNCTAD conference that declared, among other things: "We believe that a code can be produced which will make a major and positive contribution to the international transfer of technology, as well as to strengthening the technological capacity of all States, especially developing countries."

Agreement on the substance of the technology transfer issue will be affected by the efforts of Third World countries to revise the existing conventions on patents and trademarks, in particular the Paris Convention of 1883. As noted earlier, it is doubtful the Western advanced countries would consent to any changes of the Paris Convention that would weaken the basic principle of the protection of industrial and intellectual property rights. However, the possession of rights also may impose certain obligations, and these obligations could well include positive responses to obvious technological needs of the poor countries and the abstention from the abuse of patent rights.

To reach final agreement on the outstanding issues of the technology code will require compromises on both the substance and the wording of individual compromise proposals. This will depend on several factors independent of the merit of individual compromise proposals. These factors are (1) changes in policy views on the part of the participating LDC governments toward TNC activities, and by the Western countries toward the Third World; (2) sincere give-and-take attitudes, especially by Third World countries; and (3) tolerant understanding of the domestic political environment within which all governments must operate. While the implementation of the NIEO has been moved forward incrementally, the full transformation of the existing economic system

4. The Transfer of Technology 63

remains a distant and, perhaps, unreachable vista. Meanwhile, a learning process has been initiated through which the parties to the negotiations may acquire a better understanding of each other's problems and aspirations, thereby, hopefully, enhancing their inclinations to make constructive compromises.

If success were to be attained in establishing the code, its contents may have interesting effects on those LDCs, mainly Latin American, that already have national regulations in this area. To the extent that standards in the code would approximate these regulations, they would serve to bolster their legitimacy and reinforce their effectiveness. However, if standards in the code would be considerably less stringent than those in the national regulations—a distinct possibility—they could be detrimental to the legitimacy of more severe regulations in some of the countries that were leaders in the formulation of the code in the first place. Some of the Latin American countries, especially the members of the Andean Common Market, come to mind.

If agreement on this code were reached, the result in conceptional terms would be the establishment of an international regime with institutions such as the UNCTAD Secretariat and subsidiary bodies that would be assigned various functions, and with specific rights, obligations, and rules for the main actors, the UN members states, and for TNCs.[32] But even if final agreement on the code proved to be impossible, it is likely that most, if not all, provisions about which consensus was reached would be complied with by TNCs and UN member governments. In such an event, I would argue that an incipient, international regime may well also evolve, although rights, rules, and obligations of actors and participants as well as the functions of relevant UN bodies may not be fully developed or clear-cut at the beginning. As practical experience is gained with the operations of this regime, existing gaps may be filled in to the satisfaction of the participants, and a more comprehensive regime may develop.

II. Cases

5. From Dependency to Self-Reliance: an Evaluation of China's Experience of Technology Transfer

S. Ivy Lang

Conventional economic theory has been concerned primarily with technology developed in the advanced countries where technological and economic developments form an interconnected and self-reinforcing dynamic cycle. The historical process in which technology evolves is therefore put by this tradition in such a context: As wages rise, the existing technology becomes uneconomic compared with labor-saving innovations. Rising incomes allow greater investment expenditures per man. The new techniques combined with greater investment expenditure lead to rising labor productivity and, consequently, a further rise in incomes and a further incentive for innovation.[1]

Dependency theorists argue that this view of a positive impact of technology on economic development is inconsistent with the history of economic development in the less developed countries. The last three decades have witnessed a large-scale technology transfer to the LDCs from the developed countries through the sale of machinery or foreign investment, with little adaptation in the indigenous industrial sector. While this effort has made it possible for the LDCs to use the existing technology without themselves going through the difficult and costly process of developing it (the advantages of latecomers in Gerschenkron's phrase), the undesirable consequences of technological dependence have been overwhelming. The excessive reliance of LDCs upon the advanced countries for the primary source of technology in effect has distorted the pattern of development, created employment problems and uneven income distribution, and reinforced the overall dependent relationship of the less developed countries with the developed ones.[2]

The negative consequences of technological dependence in the LDCs may be explained easily by the problems that have confronted their policy makers in the process of technology transfer. First, the absence of appropriate technologies or the existence of the "suitability gap" has been identified in the literatures on appropriate and intermediate technologies as one of the major obstacles to technology transfer.[3] Many agricultural technologies, for instance, have been developed for countries with scarce labor, plentiful land, a temperate climate, and a large market of mass consumers, whereas most LDCs have different characteristics and markets. Because of the differences in economic, social, and political features between the developed countries and the LDCs, technologies devised in the advanced societies are frequently inappropriate for the latter, and adaptation with ingenuity and invention is required.

Second, the limitations of the economic system associated with unfavorable terms of trade often lead to the problem of financing the transfer of technology in the LDCs. Most LDCs export primary or lower-level manufactured goods produced by low-wage laborers in exchange for high-level manufactured goods produced by high-wage labor in the advanced countries. The unequal exchange works against the LDCs in terms of trade and allows the countries producing manufactured goods to capture gains in trade through the wage differential.[4] These unfavorable terms of trade, which lead to the shortage of foreign exchange in the LDCs, together with the enormous payments for technology and management services, have created great difficulties for the LDCs to finance the transfer of technology.

Last, but not least, political considerations of maintaining the traditional power base on the part of the LDC policy makers have been frequently in conflict with technology transfer. The introduction of foreign technologies is likely to lead to the emergence of modernizing groups that may challenge the traditional elite and reshape the traditional distribution of power. And since the primary sources of technology for the LDCs are Western democratic countries, the importation of Western technologies along with student and technician exchange programs is often associated with an influx of Western democratic ideas, which could threaten the tradiional political structure of some LDCs.

If we examine the postwar experience of most LDCs, we find that they share with one another not only the foregoing problems of development through technology transfer, but also the major objectives of development—the desire for economic growth and the quest for increasing military and political power. The Chinese case to be examined in this chapter reflects no more than the experience of a less developed country encountering all the difficulties in its process of technology transfer. What is somewhat different in the Chinese experience is its strong commitment to the principle of self-reliance and to a socialist and egalitarian pattern of development. These differences give China its peculiar distinctiveness as one of the very few countries among LDCs that have been able to break out of a dependent relationship with the developed countries.

Phases in China's Development and Their Consequences for Technology Acquisition

The evolution of Chinese development policy from its establishment in 1949 to the present has been marked by a drastic shift from dependence upon the Soviet Union to different forms of self-reliance. This sharp fluctuation between acceptance and rejection of technology from the Soviet Union can be illustrated in a quantitative way by charting the variations of China's imports of machinery and equipment from the USSR over time. The period from 1949 to 1960 is one of dependency, characterized by the heavy influx of Soviet technology. The post-1960 period, however, has witnessed a drastic decrease in the importation of Soviet machinery and equipment. These imports have remained at a very modest level since the early 1960s.

Six successive periods in China's economic development and their reflections in technology acquisition policy can be identified: (1) Reconstruction (1949–52). In its attempt to rehabilitate a war-torn economy, China depended upon the Soviet Union as its main source of technology transfer. (2) First Five-Year Plan (1953–58). Massive Soviet aid in the form of complete plants and industrial systems continued to flow into China, which facilitated an unprecedented rate of fixed capital formation. (3) Great Leap Forward (1958–60). China began to launch mass mobilization and mass participation in small-scale enterprises with "backyard" technologies. This became a transition period characterized by the continuing importation of Soviet plants on the one hand and the nurturing of self-reliance on the other. (4) Great Crisis and Readjustment (1961–65). In this period of serious economic crisis caused by the sudden withdrawal of Soviet assistance in the summer of 1960 and the failure of the Great Leap Forward, China readjusted its development strategy by giving priority to agriculture, stressing self-reliance, and resuming technology imports only on a small scale as well as from diversified sources. (5) Cultural Revolution and its aftermath (1966–76). The launching of the Cultural Revolution in 1966 marked a period of political turmoil and radical practices of a self-reliance policy by drastically curtailing the imports of foreign technology. (6) Four Modernizations (1977– the present). A new economic campaign known as the Four Modernizations program adopted a more vigorous technology transfer policy emphasizing complete plants, diversification of sources, and a more pragmatic practice of self-reliance.[5]

These different phases of China's economic development symbolize a "long march" in China's struggle for development from dependency to self-reliance. At its very beginning, the People's Republic of China did not have the foreign exchange or the economic and political contacts needed to make trade flow after the trade embargo imposed by the West during the Korean War. In February 1950 China and the Soviet Union concluded the Treaty of Friendship, Alliance, and Mutual Assistance, one of the consequences of which was a $300 million Soviet loan to pay for Soviet goods to be delivered between 1950 and 1954.[6] The value of the Sino-Soviet trade increased dramatically in this period after China acquired Soviet loans and credits to cover its trade deficits. The bulk of Soviet sales to China at this time consisted primarily of capital goods, for which the Soviets accepted agricultural goods in return.

Within three years of the founding of the People's Republic of China, the Chinese completed the basic reconstruction of their national economy and were ready for a long-term economic plan. The First Five-Year Plan covered the period from 1953 to 1957. The work of drawing up the draft plan already had begun in 1951, and, after being repeatedly supplemented and revised, was completed in February 1955, two years after the plan actually had been put into operation. The fundamental task of the plan was outlined by a Chinese official as follows: "We must center our main efforts on industrial construction; this comprises 694 above-norm construction projects, the core of which are the 156 projects which the Soviet Union is designing for us, and which will lay the preliminary groundwork for China's socialist industrialization...."[7] The emphasis on industry reflects the plan's strategy of selective growth under conditions of austerity, i.e., resources were channeled primarily into capital-intensive heavy industry that was expected to lead to rapid economic growth in the long run.

During the First Five-Year Plan period, massive Soviet aid in the form of complete plants and modern capital goods continued to be imported into China, and reliance on the Soviet Union remained the optimal policy for the Chinese leadership. In the period from 1953 to 1955 China's trade deficit with the Soviet Union continued to increase, reaching a peak in 1955 when joint stock companies and other Soviet-held assets were transferred to the Chinese (see table 5.1). Chinese purchases from the Soviet Union in this period were made possible by the $300 million Soviet loan agreed to in the treaty of 1950. This loan was exhausted in 1954. Chinese imports from the Soviet Union dropped after 1955 when the Chinest began repaying earlier loans. With the Soviet assistance, China expanded production of heavy industry from 1952 to 1959 at an annual rate of about 18 percent.[8] The massive flow of equipment and technical assistance from the Soviet Union was to provide China with the necessary technical foundation for developing a modern industrial economy.

The rapid growth of industry during the years of reconstruction and the First Five-Year Plan were not matched by growth in agriculture. By 1957 agricultural production no longer could meet increasing raw material requirements demanded by the growing industrial sector; nor could it provide an adequate food supply for the growing population and farm exports for the repayment of the Soviet loan. Increasingly concerned about the fact that agricultural insufficiency might retard further industrialization and economic growth, China's policy makers, at the meeting of the National People's Congress in February 1958, called for a "Great Leap Forward" in economic development in the next three years to replace the Second Five-Year Plan, which was to begin in 1958.[9]

This new campaign emphasized the mass mobilization of China's rural labor force to increase agricultural output and the development of small-scale native industries throughout the countryside. The new policy also called upon the Chinese to "walk on two legs," which was a plan for simultaneous development in industry and agriculture, in central and local enterprises, and in foreign technology and indigenous industry. The role of man's subjective will and effort was especially emphasized to bring human labor into full play and save on scarce capital. To make many local areas self-sufficient, local, labor-intensive industries were created throughout the country to produce a wide range of small tools for

Table 5.1. Sino-Soviet trade, 1950–79 (in millions of U.S. dollars)

Year	Total	Chinese exports	Chinese imports	Chinese trade balance
1950	325	190	135	55
1951	750	310	440	−130
1952	965	415	550	−135
1953	1,165	480	685	−205
1954	1,270	550	720	−170
1955	1,700	645	1,055	−410
1956	1,460	740	720	20
1957	1,195	745	550	195
1958	1,515	881	634	247
1959	2,055	1,100	955	145
1960	1,665	848	817	31
1961	918	551	367	184
1962	749	516	233	283
1963	600	413	187	226
1964	449	314	135	179
1965	415	225	190	35
1966	320	145	175	−30
1967	105	55	50	5
1968	95	35	60	−25
1969	55	30	25	5
1970	45	20	25	−5
1971	155	75	80	−5
1972	255	135	120	15
1973	270	135	135	0
1974	185	140	145	−5
1975	280	150	130	20
1976	420	180	240	−60
1977	340	180	160	20
1978	495	255	240	15
1979	355	165	190	−25

Source: *China Facts & Figures Annual* (Gulf Breeze, Fla. Academic International Press, 1981), p. 221.

local use so that modern, large-scale, urban-centered heavy industry could continue to receive the major share of scarce capital. What China wanted to achieve was both a high growth rate and overall balance.

The Great Leap Forward represents a transition period in China's development. The political and economic problems arising from China's dependent relationship with the Soviet Union in the previous decade had made some Chinese policy makers reevaluate the overall development policy. This transition was characterized by a continuing import of industrial raw materials, capital equipment, and, above all, complete plants from the Soviet Union, and the nurturing of self-reliance at both regional and national levels.

Several factors caused the Chinese to end most of the "Great Leap Forward" campaigns by late 1960. First, a succession of natural disasters led to bad harvests in 1959 and 1960. Food shortages appeared in 1959 and were accentuated in 1960. In the winter of 1960 malnutrition became widespread, and disease ravaged many parts of the country. Second, people's communes were rushed into action without careful investigation and prior experimentation. Managerial confusion and poor coordination caused by excessive bigness of the commune system lowered the efficiency of agricultural production and reduced the quality of the product. Third, long periods of day and night toil eventually exhausted people's enthusiasm and resulted in widespread resentment and passive resistance. Above all, the sudden withdrawal of Soviet assistance in the summer of 1960 aggravated China's economic crisis.

A new policy of "readjustment, consolidation, filling out, and raising standards" was adopted from 1961 to 1965 to cope with economic difficulties. Sectoral priorities were reversed to emphasize agriculture over light industry, with heavy industry coming last. Quality of output was emphasized as an important dimension of economic growth other than quantity and speed. Most important, the doctrine of self-reliance was set as the cornerstone of economic development and socialist construction. Self-reliance in this period did not operate to deny the continued need to borrow foreign technology. In place of an exclusive bilateral economic relationship with the Soviet Union, China began to invite many countries in Western Europe and Japan as trade partners to avoid creating a new bilateral dependent relationship. A major feature of China's technology acquisition in this period was the purchase of complete industrial installations instead of the finished means of production. Furthermore, large-scale purchases of food grains from Canada and Australia to cope with the domestic food shortages strained China's foreign exchange to the limit, and, the difficulty of obtaining long-term credits from the West forced China to resume technology imports only on a very small scale.

The convening of the enlarged Political Bureau meeting of the Central Committee of the Chinese Communist Party in May 1966 marked the launching of the Cultural Revolution on a full scale. It was initiated and led by Mao Zedong, based upon the "theory of continued revolution under the dictatorship of the proletariat."[10] In fact, the Cultural Revolution was the culmination of a long struggle between two factions in the Chinese Communist Party, fueled by divergent ideas on economic policies, foreign affairs, domestic class relations, and other issues of ideology and practice. The party cadres executing pragmatic policies in the readjustment period were purged. The coming radical mass movement led to the destruction of the party bureaucracy, sporadic economic disruption, and a decline in industrial output. Foreign trade and technology imports were disrupted, and the value of China's trade turnover drastically dropped in this period.

The Four Modernizations program was launched officially in August 1977 by the Eleventh National Congress of the Communist Party of China, seeking development and progress in science and technology, industry, agriculture, and military defense.[11] The erratic policies pursued during the span of the Cultural Revolution and its aftermath were fully replaced by a new mission of raising productivity and improving efficiency. The attainment of the goal of socialist modernization, the Chinese assumed, is a major prerequisite for the transition to full communism.

Though adhering to the principle of self-reliance, Chinese leaders see importation of foreign technology as necessary for the country's socialist modernization. In the early period of the Four Modernizations, China tried to import too much foreign technology too soon to be absorbed by the domestic industrial structure. Beginning in 1979 a readjustment was made to bring about a more balanced development in all sectors, improve the structure of economic management, and upgrade the levels of domestic education, technology and management.[12] A temporary slowdown in the rate of economic development, it was hoped, would build a more solid base for China's subsequent growth.

This review of China's history of development over the past three decades raises the following issues:

(1) What has China gained and/or lost by relying on the Soviet Union for technology imports in the fifties?

(2) By the same token, what has China gained and/or lost by basing its development upon the principle of self-reliance?

(3) What may other LDCs learn from the Chinese experience?

In the remainder of this chapter I will evaluate the Chinese experience of technology transfer in terms of, first, the general impact of foreign technology on the economic, political, and social arenas of Chinese society; second, the effects of the imported technology upon China's indigenous development of technology; and last, the transferability of the Chinese experience of technology to other LDCs.

China's Technological Dependence upon the Soviet Union: A Cost-Benefit Analysis

Although "dependency" was originally adopted by students in Latin American studies to refer to the highly asymmetrical relationship between European and Latin American countries in the historical process of their integration into the international capitalist system, many social scientists have argued recently that it should have wider applicability.[13] To put it in a broader context, dependency refers to any situation in which the economy of one country is conditioned by the needs and interests of another country, or, decisions made in the dominant state determine the pace, direction, and content of economic growth and development in the subordinate state.[14]

Dependency has had three historic forms: colonial, financial-industrial, and technological-industrial. The first two forms were trade-oriented; precious metals, agricultural products, and raw materials were brought to the center states from the periphery in exchange for manufactured goods. The characteristic of the present form of dependency, however, is the establishment of manufacturing facilities by multinational corporations in most Third World countries. These facilities produce both for the internal market and the integrated world market of the corporation; in other words, dependency in modern form is a technological dependence more than a trade dependence.[15]

The generally shared desire for industrialization among LDCs today requires them to acquire technology through either its importation from developed countries or by means of investment by multinational corporations. The almost exclusive reliance on foreign sources for technology has become a universal phenomenon among the LDCs. Technological dependence usually leads to such undesirable consequences as "inappropriate technology transfer." It also is said to foster international and local inequality, unemployment, regional and social polarization, foreign indebtedness, and uneven economic development.[16]

China's policy of technology transfer in the fifties was one of dependency. The imbalance in technological capability between China and the Soviet Union can be indicated by the large-scale flow of Soviet machinery and technological assistance into China. For example, more than 30 percent of Soviet exports to China during the First Five-Year Plan were in machinery; of the 51,500 machine tools produced during the plan, 43,500 required Soviet blueprints; from 1949 to 1955, three thousand Soviet technical books were published, with a total printing of twenty million copies; some twenty thousand students, technicians, and scientists received advanced training or education in the Soviet Union between 1951 and 1962; more than ten thousand highly skilled Soviet technical specialists were sent to China between 1950 and 1960.[17] On the other hand, during the same period China furnished the Soviet Union with more than 1,400 million new rubles' worth of mineral products and metals, and 2,100 million new rubles' worth of grain, edible oils, and other foodstuffs. Among the most important items were 5,760,000 tons of soybeans, 2,940,000 tons of rice, 1,090,000 tons of edible oils, and 900,000 tons of meat.[18]

The massive technology transfer from the Soviet Union that occurred in the fifties clearly entailed significant benefits. First, in the period following the establishment of the People's Republic, the economic and technological base of China after twenty years of civil and international warfare was so weak that China desperately needed foreign assistance. The American response to the Communist Revolution in China was so hostile that in practice the Soviet Union was the only country to which the Chinese could turn. Soviet technological assistance helped China reconstruct its war-torn economy and strengthen its national security. Second, while China exported primary goods and imported industrial goods, the industrial products were mostly capital goods instead of durable consumer goods. Capital goods did, in the long run, give China the capacity to produce independently both capital and consumer goods. From 1952 to 1959 China's growth rate was as high as 17.6 percent per year, compared with 6.6 percent in India.[19] Finally, between 1954 and 1959 the Soviet Union provided China with more than 24,000 complete sets of scientific and technical documents, including 1,400 designs for major enterprises.[20] All this documentation was worth billions of dollars and was given when nowhere else would China have been able to obtain such scientific, technical documentation in the volume necessary for extensive economic construction.

China was not unaware of the negative impacts of technology transfer from the developed countries, but it distinguished between socialist aid and imperialist aid from the capitalist powers. Its understanding of Socialist aid can be summarized by Article 5 of the Treaty of Friendship, Alliance, and Mutual Assistance signed by China and the Soviet Union in 1950: economic assistance must be "in conformity with the principles of equality, sovereignty, territorial integrity and noninterference in internal affairs" of the other party.[21]

The Chinese preoccupation with socialist aid as something promoting international equality caused a lack of recognition among Chinese leaders of the possible costs of their policy of technology transfer. After 1958, however, growing consciousness of the costs of the technological dependence upon the Soviet Union was reflected in internal party conflicts between the Maoists and cadres educated in the Soviet Union, which culminated in the "campaign against experts" and the "backyard furnace movement" led by the Maoists during the Great Leap Forward period.

The costs entailed by the excessive dependence upon the Soviet Union for technology transfer are several. First, from 1955 to 1956 China began to repay the debts incurred by the massive influx of Soviet technology through substantially increasing their exports to the Soviet Union. During this period they became increasingly conscious of the adverse terms of Soviet credit and trade. This form of indirect economic exploitation can be illustrated by the fact that many goods China imported from the Soviet Union were priced much higher than those on the world market, and that the Soviet Union insisted on collecting debit service installments from China even in 1958 to 1959 when China's need to finance its imports from the Soviet Union strained Chinese foreign exchange resources to the limit.[22]

Second, any development strategy should take into consideration the nation's so-called factor endowments. In a country such as China, where capital is relatively scarce and therefore expensive, while labor is relatively abundant and therefore cheap, massive imports of machinery and plants from the Soviet Union under the Soviet-imposed development policies adversely influenced China's allocation of capital resources between agriculture and industry and led to an uneven development pattern stressing heavy industry at the expense of agriculture. Overtaxing agriculture resulted by 1956–57 in a reduction of agricultural production and of raw materials to meet the demands of the industrial sector. The insufficiency of agricultural production finally led to the slowdown of economic growth and hindered further industrialization in the late fifties.

Third, interest in Communist solidarity in the early fifties was sufficiently powerful to restrain China from asserting national interests. In the years immediately after the founding of PRC, Chinese leaders granted the Soviet Union special privileges in Manchuria, tolerated the Soviet plunder of Manchurian industry, and accepted a system of joint commercial enterprises.[23] The Soviet Union, in the name of the unity of the international Communist movement, tried constantly to influence the policy-making process in China. Since party leaders and cadres had established a high degree of autonomy in their long years of revolution and civil war, by late 1957 they became aware that Soviet technological assistance in the name of fraternal comradeship had in effect created a status distinction between China and the Soviet Union that was based by no means on equality, reciprocal benefit, mutual respect, and noninterference.

Finally, according to Galtung's structural theory of imperialism, economic dependency in the Third World tends to transform the local elite into a client social class. While the indigenous ruling class (center of the periphery) possesses political sovereignty, it performs important functions on behalf of the interests of the elites in the dominant country (center of the center). In exchange, the foreign elites insure wealth, power, and other privileges to their local clients.[24] Since China's relationship with the Soviet Union was based more on politics and ideology than on corporate interests, the issue of co-optation actually revolved around preferences for development strategy. Chinese intellectuals, scientists, and technocrats who had been sent to the Soviet Union for advanced training or who had worked closely with Soviet personnel in China were therefore most likely to be co-opted; they vocally advocated their belief that Soviet methods, attitudes, and goals were correct and superior with respect to China's development, with a concomitant neglect of the country's indigenous conditions.

China's policy of technology transfer in the fifties, as discussed above, entailed both benefits and costs. It is not easy to deny that Soviet technological assistance not only helped China pass its most difficult time in the early days of its establishment, but also laid the foundation of industrialization during subsequent years. However, the policy allowing for excessive dependence upon the Soviet Union for technology imports if not stopped in time would have aggravated the already beginning tendency of draining foreign exchange resources, uneven growth, satellite status, and the emergence of a domestic co-opted class. The causes of the Sino-Soviet rift in the early sixties are several, yet the growing awareness of the costs entailed by the old policy and the resolution to end the dependency relationship was one of the most important.

Technology Transfer Based upon the Principle of Self-Reliance

The 1960s witnessed a drastic shift in development strategy from dependency to self-reliance in China. The sudden withdrawal of Soviet assistance in the summer of 1960 combined with successive crop failures brought the Chinese economy to a state of crisis. In the long run, however, the new development policy based upon the principle of self-reliance that was followed after 1960 has enabled China to achieve a degree of technical and economic independence.

Self-reliance, as a development objective, encompasses several concepts and

Table 5.2. China's import dependence compared with that of slected countries and the world, 1975

	GNP in millions of U.S.$	Total imports in millions of U.S.$	Import dependence percentage
World	6,130,000	903,200	14.70
China	299,000	7,385	2.46
Unites States	1,498,000	102,984	6.87
Soviet Union	865,000	36,969	4.27
Japan	484,000	57,881	11.95
West Germany	422,000	74,208	17.58
France	340,000	54,247	15.95
Brazil	97,800	13,658	13.96
India	85,300	6,382	7.16
Finland	28,000	7,607	27.16
Nigeria	25,000	6,041	24.16
South Korea	18,700	7,274	38.89
Taiwan	14,400	5,959	11.38
Hong Kong	7,000	6,767	96.67
Albania	1,200	160	13.00

Source: *United Nations Statistical Yearbook, 1976*; *U.S. Government Handbook of Economic Statistics* (1976).

traits, with the emphasis given to each of these changing from time to time. As an operational policy it has been applied in several different contexts and for different purposes. By and large, China's self-reliance policy can be identified by the following indicators. First, self-reliance relates to a low degree of import dependence. Quantitatively, China's reliance on imports is marginal. Imports are necessarily a small sector relative to GNP in any large continental or subcontinental economy. China's import dependence index is low even in comparison with other large countries (see table 5.2).

Second, self-reliance relates to a low degree of partner concentration in trade. China's import values as a percentage of GNP have shown little variation from 1950 to 1976 (the lowest is 0.95 percent in 1969; the highest 2.61 percent in 1974) after the shift of policy to self-reliance.[25] The difference, however, is reflected by the direction of trade. The Soviet Union was China's major trade partner in the fifties. Trade with the Soviet Union used to account for about half of Chinese trade, but the figure dropped drastically after 1960.[26] Self-reliance was reflected by the attempt to involve multiple trade partners so as to avoid repeating any kind of dependency relationship. The number and variety of China's trade partners since 1960 has gradually increased to include Japan, Western European countries, northern American countries, Australia, Southeast Asian countries, Latin American countries, Hong Kong and Macao, Eastern European countries, and the Soviet Union.

Third, self-reliance relates to a low degree of financial dependence. China's self-reliance is most pronounced when gauged by the criterion of international financial dependence. China incurred sizeable trade deficits in the 1950s, amounting to a total of about $1.2 billion. These were financed by Soviet credits, which were beginning to be repaid in 1955. In 1956 this situation was reversed, and since then and until 1972 China accumulated trade surpluses totaling about $2.3 billion, which were used to repay the Soviet aid and finance China's foreign aid program. Therefore, until recently China has been a net exporter of capital, a most unusual situation for a less developed country and particularly for a country experiencing rapid economic growth.[27] As a result, China is one of the very few LDCs free of any long-term debt. However, in the early phase of the Four Modernizations, as a result of an ambitious import program combined with world recession-induced export difficulties, China incurred large trade deficits that are being financed by a combination of short-term and intermediate-term credits. These balance of payments difficulties were reversed by slowing down the importation of heavy industrial equipment and agricultural goods beginning in 1979.

Finally, self-reliance relates to a low degree of trade commodity concentration and to an independent foreign policy. A developing country like China could, if it wished, follow an open foreign trade orientation based on the principle of comparative advantage advocated by some economists. That is, it could specialize in the production of certain types of farm products for which growing conditions are particularly favorable, use these both for home consumption and exports, and rely on imports to meet a significant share of food supply requirements. While China imports some foodstuffs, these are marginal.[28] In the Chinese view, heavy reliance on imported farm products would expose the country to the risk of sudden supply embargoes, rendering it vulnerable to foreign pressures and violating the principle of self-reliance.

Self-reliance as a policy of technology transfer in China has two major themes: the importation of technology based on favorable terms, and integration and cooperation between imported technology and domestic enterprises. Since the Chinese recognized that imported capital goods can play a significant role as a component in domestic investment, the result of considerations of self-reliance has not been an abandonment of technology acquisition from abroad. They also have recognized that the essential problem was not whether to import technology and know-how but how to manage the import policy so as to fit with China's preferred economic and political arrangements, including the aim of self-reliance. In this light, technology transfer based upon the principle of self-reliance since 1960 has been a concern of selectivity, of neutralizing undesirable side effects, and of the domestication of imported technology.

To obtain favorable terms on imported technology, the Chinese authorities have been careful not to become overly dependent on one or a few suppliers, and they also tried not merely to import machinery and finished capital goods, but to obtain blueprints, specifications, laboratory reports, and to dispatch technical delegations abroad. The two most significant forms of imported technology in China today are prototypes for learning and copying as well as complete production plants and processing equipment. China procured a wide variety of prototypes at bargain prices by purchasing display models of the most advanced equipment at the conclusion of industrial fairs and technology exhibitions held in China by developed countries. The prototypes were used to strengthen the Chinese manufacturing process by means of "reverse engineering," i.e., taking the imported machinery apart and reconstructing it. Furthermore, a far more rewarding but also more expensive form of technology transfer is the turnkey production plant or comprehensive equipment package, complete with the technical data and advisory assistance to set it up. Importing such modern production plants has offered a relatively quick solution to the nation's economic problems, but its costs and maintenance create formidable problems.[29]

More important, self-reliance in the context of technology transfer policy is reflected by China's effort to facilitate integration and cooperation among its three major types of industrial enterprises: the large plants initiated by Soviet-assisted projects, the small and medium urban enterprises, and the expanded rural industries. The Soviet-aided industrial plants, which were expected to form the core of China's modern industrial sector, were found in many cases to be too large and too specialized for China. The plan that sought to achieve major structural changes in a short time would have been realized without integrating the imported technology into the existing indigenous foundation of industry and without the cooperation among three major types of industrial enterprise.

Foreign technology flows into the most advanced plants and activities, frequently the large-scale central plants. The medium-scale provincial and municipal plants and small-scale rural industries derive their technological advancement from a "trickle down" process of internal diffusion from higher to lower economic and administrative levels. In such a process, all levels of industry have been able to upgrade their technical competence through imported technology.

The small and medium urban enterprises, often part of the heritage of pre-1949 industry, are largely responsible for the flexibility, innovative success, and responsiveness to demand by providing technical change drawn from both foreign and domestic sources. Their ability to reorganize human capital and to respond promptly to changing demands proved to be important for providing an essential complement to the large but often inflexible Soviet-assisted enterprise, and for regaining economic momentum following the crisis of the early 1960s. Their contributions are best illustrated by the growing outputs of China's petroleum and chemical fertilizer industries after 1960.

Small rural plants served to support other units of enterprise by meeting local demands, reducing the production burden of leading units, and thereby allowing advanced industrial units to concentrate on innovation and technology advance. The development of rural industries also conformed with China's values of industrial dispersion for the purpose of national security, of promoting labor-intensive manufacturing methods, and of cooperation between industry and agriculture. Five major rural industries, fertilizer, cement, farm machinery, energy (coal and hydropower), and iron and steel, for instance, all have great impacts on farming.[30]

Chinese leaders are well aware of the costs they have paid for dependency avoidance and self-reliance—a slower economic growth rate, among others. However, China now has made significant headway in building up a capacity to selectively borrow foreign technology and to integrate the imported technology into a dynamic and developing economy under Chinese control. The costs paid for breaking technological dependence, it is believed, are clearly outweighed by the benefits.

China as a Model

China's policy of technology transfer based upon the principle of self-reliance is successful in the sense that it has made it possible for China to become probably the only LDC in the world today that has achieved an independent position in not only domestic affairs but also in international relations. For a country that has been humiliated and exploited by the West, Japan, and the Soviet Union, its political achievement is enormous; for a country that used to be unable to feed its large population, its economic achievement is amazing; for a country that began to strive for a transformation from a semifeudal and semicolonial system to a socialist system only three decades ago, its social achievement is undeniable. China's self-reliance policy in technology transfer is far from perfect, as demonstrated by the "radical" programs of the Cultural Revolution or the "aggressive" programs in the early phase of the Four Modernizations. Nevertheless, its determination to make readjustments at any expense to stick to its commitment to self-reliance is what has made the Chinese experience a distinctive model.

To what extent is the Chinese self-reliance model transferable? Are there any lessons to be drawn from the Chinese experience that could be applied in formulating technology transfer policies for other LDCs? The notion of transferability has been explored in two ways by previous studies. One approach considers that it is much easier for a large country like China to pursue a policy of self-reliance in technology development without sacrificing economies of scale or efficiency. This option is not open to small economies that, given their limited resources and internal markets, must necessarily take advantage of the international division of labor if they are to develop. Another view suggests that the Chinese self-reliance model is deeply imbedded in its economic, political, and social system as a whole, so that before borrowing from the Chinese experience LDCs must first carry out a Communist revolution.

It is argued here that the Chinese self-reliance model in technology transfer is transferable not only to small countries, but also to private enterprise—oriented market economies. Self-reliance does not mean reducing a nation's involvement in international technology transfer, but rather in maximizing the degree of national control over the imported technology and the integration of the imported technology into the indigenous industrial foundation. Therefore, it should be equally transferable to both large and small countries. Also, as long as a country clearly recognizes the danger involved in the mechanical absorption of foreign material, it will have the need and desire for adopting the self-reliance model in technology transfer regardless of its social system.

6. Nuclear Power in the Philippines: Technological Choice and Dependence[1]

Robert L. Youngblood and Melba D. Solidum

Consider the following hypothetical situation: You are the president of a small, developing island nation with no domestic fossil fuel energy supplies to encourage rapid industrialization. You are approached by several multinational corporations interested in selling your nation a nuclear power plant. Nuclear power, they tell you, will supply the electricity essential to your industrialization plans. It is cheap and clean. Appealing to your sense of pride, they point out that nuclear power is a symbol of an advanced society. One company makes you an extremely good offer on a nuclear power complex, and, unable to resist their sales pitch, you agree to purchase not one, but two plants.

But problems begin to develop. The price of the two reactors inexplicably (and inexorably) begins to rise, first by 57 percent within one year, and then, over the following year, by another 83 percent for a single plant! Your nation is locked in a volcanic, earthquake-ridden part of the world, and the site chosen for the plant, some five miles from a potentially active volcano, is discovered to be subject to earthquakes that may exceed the safety limits of the reactor. Even more troubling, you discover that the plant you have purchased is of obsolete design and, to add insult to injury, could end up costing more than reactors of similar size under construction elsewhere.[2]

Because the Philippines are a former American colony, the country's decision in 1973 to purchase a nuclear reactor from Westinghouse Corporation is an interesting case in terms of current theories of dependency, especially Johan Galtung's model of structural imperialism. According to Galtung, imperialism is a special kind of dominance relation between a Center and a Periphery nation such that there is a harmony of interest between the center of the Center (cC) and the center of the Periphery (cP), and this harmony of interest becomes a bridgehead of the Center nation into the Periphery nation, resulting in an asymmetrical relationship that tends to keep the Periphery nation in a dependent status. At the same time, there is an inherent disharmony of interests between center and periphery in both the Center and Periphery nations and a disharmony of interests between the periphery of the Center nation (pC) and the periphery of the Periphery nation (pP). Galtung maintains that formalized imperial/dependent relationships exist when:

(1) there is *harmony of interest* between the *center in the Center* nation and the *center in the Periphery* nation.

(2) there is more *disharmony of interest* within the Periphery nation than within the Center nation,

(3) there is disharmony of interest *between the periphery in the Center nation* and the *periphery in the Periphery* nation.[3]

A harmony of interest between foreign and Filipino elites not only characterized the Spanish and American colonial periods, but also has been a major factor in Philippine politics and economics since independence in 1946. In return for rehabilitation aid under the terms of the Bell Trade Act of 1946, for example, Filipino leaders pushed through an amendment to the 1935 Constitution allowing Americans "parity" with Philippine citizens in owning and operating public utilities and in developing natural resources. In 1947 they signed a Military Bases Agreement that gave the United States a ninety-nine year lease on twenty-three bases, including Clark Air Base and Subic Naval Base. And with the declaration of martial law in 1972, President Marcos defeated a number of anti-American proposals before the Constitutional Convention and reversed several Supreme Court decisions opposed by the American business community. Advised by a group of American-trained technocrats, the President outlined an economic development program that included opportunities for considerable American and other foreign investment.

To be sure, economic modernization in the Philippines requires additional energy installations, but the decision to invest in a nuclear reactor rather than develop hydroelectric, geothermal, and coal resources or to investigate other energy options must be evaluated in the context of linkages between the Philippine elite and the United States and a development strategy that is capital-intensive and relies on the transfer of foreign technology. Some of the negative consequences of the Philippines' development strategy are becoming increasingly evident, including a foreign debt of over $15 billion, the stalling of a number of major industrial projects, increasing government equity in unprofitable luxury hotels, and recent criticism by the World Bank of the regime's handling of the economy. With the gap between rich and poor growing, the government announced in 1981 the inauguration of a rural entrepreneurship program called the Kilusang Kabuhayan at Kaunlaran (KKK—National Livelihood and Progress Movement). Yet investment in the program is hampered by previous commitments to other projects such as the $1.9 billion nuclear reactor. The decision to purchase a nuclear reactor must be evaluated within this context, for, clearly, technological choices have ramifications for the economy that affect the well-being of millions of Filipinos.

In analyzing the Philippine nuclear project we will examine the key elites involved in the purchase decision, the various groups and issues surrounding an opposition to the project, the government's response to critics of the nuclear power plant, and, in terms of Galtung's imperialism/dependency model, the various implications of the nuclear technology choice, including:

(1) The greatest demand for increased energy comes from an industrial sector heavily influenced by foreign investors.

(2) The nuclear power plant's electricity will be inequitably distributed within the country.

(3) Construction of the plant will not lead to the democratization of social and economic opportunities.

(4) The construction of the nuclear power plant reflects a harmony of interests between local elites (center of the Periphery) and foreign elites (center of the Center) and will not lead to the goal of ultimate self-reliance.

Purchase of the Plant

The Philippine government's decision to build the Philippine Nuclear Power Plant, to be in operation by 1985 and expected to supply less than 7 percent of the country's future energy needs, was made in July 1973 through a presidential decree, three months prior to the increase in world oil prices.[4] At that time there were plans to build eleven nuclear reactors at three sites by the year 2000; by 1976 construction plans had to be cut drastically due to severe inflation and oil price increases that reduced hopes for heavy capital investments. A report published in 1977 by two members of the Philippine Atomic Energy Commission (PAEC) voiced concern over critical issues such as cost, uranium supply, and waste disposal, which contributed to a decision to construct two reactors at a single site. And by March 1981 Energy Minister Geronimo Velasco ruled out the purchase of the second plant in favor of developing "recent discoveries of new geothermal and coal sources of energy."[5]

The nuclear power plant under study is a 620 MW, two-loop, pressurized, light-water reactor purchased from Westinghouse Corporation and currently under construction on a 300-hectare site in Morong, Bataan, forty-five miles west of Manila.

Who Was Involved?

The Export-Import Bank (Eximbank), a United States government agency that provides financial subsidies for establishing overseas markets, supported the $1.1 billion (1975 prices) nuclear plant project by extending a $644.4 million loan, of which $277.2 million was in loans and $367.2 million in loan and bond guarantees for the Philippine National Power Corporation (NPC), a foreign public corporation. The Eximbank also gave the Philippine government an extra six years beyond the normal eleven-year grace period in which to begin payments, meaning that since financing was approved in December 1975 the payments will not start until 1992.[6]

Westinghouse Corporation won the contract to build the nuclear power plant despite lower bids and better terms from General Electric and, reportedly, from French and West German firms. Before the Eximbank committed its financial support, for example, General Electric offered to construct two 600 MW reactors for $700 million; Westinghouse's counteroffer was $500 million for both. But by June 1974, "after the Marcos government and Westinghouse signed a letter of intent," the prices went up to $1.2 billion for two reactors, and in November 1975 the cost jumped to $1.1 billion for one and $1.6 billion for two. Thus, in less than two years the prices more than doubled for each facility, while, ironically, during the same period Westinghouse offered Spain a 930 MW reactor for only $687 million. Moreover, U.S. financing of the Philippine nuclear project led to allegations that the Eximbank was bailing Westinghouse out of financial difficulties resulting from slumping demand in the nuclear industry. It also is significant that at the same time Westinghouse was embroiled in a $2 billion suit for having failed to supply uranium at agreed-upon prices to several customers.[7]

There are other instances of irregularities that seem to point to collusion between Westinghouse and the Eximbank. Consider the following:

> In June 1974, while the Philippine government was still debating whether to purchase a plant from General Electric or Westinghouse, a Philippine team arrived in Washington to discuss financing with Eximbank. The team was shocked to find Westinghouse representatives sitting in the same room with them. "It gave you a feeling a deal had been made," one member later said.
>
> Eximbank made its commitment to finance the purchase in Oct. 1974 without seeing Westinghouse's formal bid. This was not presented until March 1975.
>
> Then U.S. Ambassador William Sullivan had an inordinate interest in the transaction. He cabled Eximbank that, "The Embassy consider(s) a good deal of American prestige riding on the Westinghouse deal and . . . we intend to watch the project closely."
>
> Eximbank guaranteed $367.2 million in Philippine government bonds. This is the first time the Bank has ever guaranteed the bonds of a foreign corporation sold on the U.S. market.
>
> Eximbank never questioned the fact that the price quoted by Westinghouse for the reactor increased by over 400 percent between 1974 and 1975.
>
> Eximbank allowed the loan funds to be disbursed before Washington had received an export license for the plant from the NRC.[8]

The go-between responsible for the success of the Westinghouse contract was Herminio Disini, a cousin-in-law and regular golfing companion of President Marcos and head of the Herdis Management and Investment Corporation. Disini, through Herdis, acquired Asia Industries Incorporated, formerly the distributor of Westinghouse products in the Philippines, which became a major subcontractor for the project. Without any bids the construction contract was awarded to Power Contractors, Inc., a consortium of Asia Industries. Other Herdis subsidiaries involved in the nuclear project include Technosphere Consultants Group, which provides the engineering and construction management, and Summa Insurance Corporation, which underwrites the $693 million insurance package for the plant. Summa's selection as insurer was interesting given that it ranked 89th among 106 Philippine insurance companies and earned only $17,350 in 1976.[9]

Allegations have surfaced regarding the payment of an estimated $4 million to $35 million "commission" to Disini for facilitating the deal. Although a spokesman for Westinghouse averred that "to the best of our knowledge there have been no improper payments," the corporation evidently admitted to the U.S. Securities and Exchange Commission in 1976 the existence of "questionable payments," "commission payments," and "improper allowances" in connection with overseas projects.[10] Significantly, following allegations of corruption, President Marcos ordered Disini divested of some of his companies, yet no public investigation of the charges was held. Disini maintained he was innocent and the object of unwarranted attacks by political opponents of the regime.

Opposition to the Nuclear Project

Domestic and Foreign Groups

The Philippine decision to construct a nuclear power plant has not been without opposition. Domestically, organizations such as the Kilusan para sa Kapangalagaan ng Kapaligiran (Movement for Environmental Protection), Kilusan ng mga Nagmamalasakit na Mamamayan ng Bataan (Movement of Concerned Citizens of Bataan), and the Kapisanan ng mga Kimiko sa Philipinas (Philippine Chemists Society) criticized the project and appealed to Marcos to suspend construction and to conduct an exhaustive review of the government's nuclear program. A group of concerned citizens engaged the PAEC in a dialogue on nuclear power in March 1978, while other groups held special prayer services in conjunction with worldwide demonstrations against the Philippine nuclear plant in April 1978.[11] Similarly, a gathering of two hundred public schoolteachers, nuns, and priests from three towns within a radius of sixty miles from the plant expressed concern about living near the facility, and Assemblyman Antonio Roman of Bataan spoke on the floor of the Batasang Pambansa (National Assembly) against construction of the plant. Following antinuclear moves by such organizations as the National Research Council of the Philippines, the Biology Teachers Association, and the Bataan Mayors' League, the Bataan Sangguniang Panlalawigan (Provincial Council) passed a resolution officially opposing the nuclear project.[12]

But by far the stiffest opposition to the plant was led by former senators Lorenzo Tañada and Francisco "Soc" Rodrigo and by Joker Arroyo, a prominent Manila lawyer. The major accident at Three Mile Island plus a letter by Senator Tañada to President Marcos on June 14, 1979, resulted in a suspension of construction the following day pending an inquiry into the plant's safety features, while Executive Order No. 539 provided for the creation of an investigation committee, headed by Assemblyman Ricardo Puno, a member of Marcos' political party, the Kilusan ng Bagong Lipunan (KBL), charged with holding public hearings on the project and with analyzing the technical aspects of the plant's safety equipment. The Puno Commission held fifty sessions from June 23 to September 14, 1979. Tañada and Arroyo, however, walked out near the end of the public hearings, arguing that in rushing the proceedings the commission had "in effect . . . prejudged the issue."[13] Even as the hearings were still going on, the Puno Commission recommended a return-to-work order with concurrence from Malacañang (the Presidential Palace).[14] Yet at the same time the commission concluded:

1. The Bataan Nuclear Plant as designed is not safe. . . . Thus, it is a potential hazard to the health and safety of the public.
2. The Bataan Nuclear Plant design needs fundamental changes and additional safeguards. It appears that Westinghouse nuclear reactors do not have, among others, reliable emergency core-cooling systems.
3. The frequency of accidents in nuclear power reactor plants, of which the Commission takes public notice, is an ominous sign that safety is not adequately assured. It is imperative that the requisite safety devices be installed if it is decided to continue with the nuclear plant.
4. The crucial problem of nuclear waste disposal has not been solved, as the Inter Agency Committee charged with finding a final repository for these wastes has yet to locate a suitable place.[15]

Opposition to the project from abroad also abounds. Congressman Clarence Long, for example, in U.S. congressional hearings, sharply criticized the State Department and the Eximbank for guaranteeing the $644.4 million loan without first investigating the nature of the project. In separate letters to Secretary of State Cyrus Vance and Nuclear Regulatory Commission (NRC) Chairman Joseph Hendrie, Long strongly argued against granting the export license for the power plant on the grounds that it was uneconomic, unsafe, and environmentally unsound. He also stressed that the project represented a conflict of interest.[16] Some of Long's concerns were echoed by Daniel Ford, executive director of the Union of Concerned Scientists, in a critical report to Marcos in February 1979, and by Frank von Hippel, chairman of the American Federation of Scientists, in a letter to the Puno Commission in August 1979. Ford emphasized unresolved technical problems in the plant's design, while von Hippel suggested a group of non-nuclear scientists review the site and safety designs of the Bataan plant.[17]

At a grass-roots level, demonstrations were held in San Francisco, Washington, D.C., Canberra, Sydney, Melbourne, Tokyo, Stockholm, Rome, Amsterdam, Utrecht, and London on the International Day of Protest Against Nuclear Reactor Exports to the Philippines, April 27, 1978. The event was sponsored by the Coalition for a Nuclear-Free Philippines, composed of the Friends of the Earth, People Against Nuclear Power, and U.S.-based anti-Marcos groups such as the Friends of the Filipino People (FFP), the Anti-Martial Law Coalition (AMLC), and the International Association of Filipino Patriots (IAFP).[18] Other efforts abroad included a visit to Prime Minister Trudeau by members of the Committee Against Nuclear Pollution in the Philippines to prevent the Canadian sale of uranium to the Philippines. Groups in Australia opposed to the sale of uranium to the Philippines report aborigines were pressured into allowing the federal government to mine yellow cake under a threat of losing tribal land rights.[19]

Yet, despite the opposition of antinuclear groups in the Philippines and abroad, the NRC, by a vote of three-to-one with one abstention, granted Westinghouse an export license for the nuclear plant in May 1980—nine months after Westinghouse filed suit against the commission for delaying the issuance of the license.[20]

Issues

The biggest controversy regarding the Bataan plant centered on the issue of safety—first, in terms of design, engineering, and technology, and, second, in terms of the site and location. The plant worries many nuclear engineers and scientists because research has demonstrated that the design lacks necessary safeguards and is unable to pass stringent requirements of plants built in the United States. Part of the problem is the fact that the design of the Philippine plant was modeled after a two-loop Yugoslavian plant under construction since 1974, which, in turn, was modeled after a two-loop Brazilian plant, which, in turn, was modeled after a similarly designed plant in Puerto Rico whose construction was terminated in 1972, resulting in the design avoiding a comprehensive review by the NRC.[21] Additionally, according to Robert Pollard, a nuclear safety engineer of the Union of Concerned Scientists, there are many unresolved design problems. Although the Ministry of Energy maintains that the NPC had to meet 146 safety requirements stipulated by the NRC, the statement is misleading since 102 of these should have been incorporated in the original contract. Of the other forty-four requirements, only twenty were mentioned in the renewed contract. Moreover, a number of stipulations required as a result of the Three Mile Island disaster have not been included, and, as Pollard pointed out, in at least one case "some of the requirements concerning steam generator corrosion are discussed only for the purpose of saying they will not be met."[22]

That the safety design problems must be faced directly is underscored by numerous plant failures and accidents elsewhere in the world—even since the major breakdown at Three Mile Island in 1979. For example, in January and March 1981, cracked pipes at the Tsuruga Plant in Japan, 225 miles west of Tokyo, exposed 101 workers to highly radioactive waste, resulting in the closing of the facility; while in September 1981, inspections revealed that the steel shells surrounding the uranium cores of thirteen reactors, including the Connecticut Yankee, were becoming brittle.[23] Since the plants are built around the shells, the shells are irreplaceable. Additional reports indicate that sixteen other plants, including New York's Consolidated Edison Indian Point 2 plant, have rusting steam generators, which, while not as serious as brittle reactor shells, may necessitate a number of plant closings. At least thirty-six of the forty-eight steam generator plants in the United States have problems with corroded tubes.[24] More recently, in January and February 1982, cooling system leaks in other U.S. reactors have caused expensive shutdowns.[25]

Questions also have been raised about the technical expertise available to run and maintain the Philippine plant. Pollard points out, for example, that the "PAEC describes as 'normally open' exactly the same valve which is in fact normally closed but stuck open during the TMI accident"; and according to the PAEC these valves are "'connected in series' with other valves, when in reality, the valves are, and must be, connected in parallel." He also argues, as have a number of Filipino scientists, that the Philippines has neither the technology for the input (uranium fuel) nor the output (waste disposal).[26] These problems are compounded by the PAEC's lack of "necessary depth of technical expertise and experience" to monitor the plant effectively. Perhaps more fundamentally, Congressman Long earlier questioned whether Power Contractors, a newly organized construction firm, had the experience to build the plant according to safety specifications.[27]

The other aspect of the safety issue has to do with the location of the nuclear plant. Despite reassuring pronouncements by the NPC and Ebasco Services, Inc. (the NPC consultant on site investigation), the Bureau of Mines stressed that the "plant is situated near several earth fracture lines," while the Commission on Volcanology and the International Atomic Energy Agency confirmed the potential hazards resulting from volcanic activity, such as lava flows, mud slides, and ash falls, in the area. The nuclear plant is surrounded by five volcanos, namely Mt. Natib (active, ten miles away), Mt. Taal, Mt. Banakoa, Mt. San Cristobal (all active, between sixty and ninety miles away), and Mt. Mariveles (inactive, twelve miles away). According to volcanologists, problems and dangers can come not only from the main craters but also from "parasite craters."[28] David Leeds points out, for example, that the plant should be able to withstand earthquakes of an intensity of point 8.0 on the Richter Scale. Leeds notes that the facility is inadequately designed to withstand such earthquakes and volcanic eruptions, arguing the "risks are not adequately stated," "the seismic criteria are too low," and "nearby volcanic activity has not been faced" forthrightly.[29] That the plant is "located less than six miles from the entrance to the Subic Bay Naval Base, where the U.S. Navy's Seventh Fleet stores 110 million gallons of diesel oil, 1.7 million gallons of jet fuel and ammunition for 20 surface warships, two aircraft carriers and the 200 warplanes assigned to the two carriers" also adds to concerns over the project site.[30]

A second set of issues surrounding the site location deals with the environmental effects of the project. Farmers, fishermen, and environmentalists alike anticipate negative consequences from the plant's water discharge on marine life. The NPC revealed that discharged seawater used to cool the plant will be "five to seven degrees hotter than the surrounding water" and will significantly raise the temperature of "one kilometer of sea" from the Napot Point release area with a potential for causing thermal shock to many sea organisms and driving fish from the area.[31] This is especially important in the Bataan area since the milkfish industry is well developed and many fishermen stand to lose their livelihood just as numerous farmers lost their land, with little hope of receiving just compensation, to the construction site.[32]

A critical issue that every country with a nuclear program faces is the problem of waste disposal. The wastes remain radioactive and toxic up to 250 thousand years, and there are no guarantees that any currently used storage method will contain nuclear wastes in perfect isolation indefinitely. The Philippines has considered burying its wastes below ground, in granite rock in Northern Palawan, but given the country is situated in one of the world's most active earthquake zones, this option is discomforting. The problem is even more serious in light of the fact that nations like Japan and the United States, with advanced technology, are still searching for solutions to nuclear waste disposal problems. Japan even tried to convince Guamanians and Micronesians that an experimental dumping of low-level radioactive wastes, presently stored in metal drums at reactor sites, in the Pacific island area was safe. Predictably, the plan met vehement opposition from the islanders.[33] The United States deposits low-level radioactive wastes in dumps and high-level military wastes in remote sites, but no disposal program has yet solved the long-term environmental contamination problems. And deterioration of containers and leakages already have caused serious problems at some sites.[34]

Another major problem associated with nuclear programs is economic. The plant itself is extremely capital intensive—a severe problem for Third World countries like the Philippines with huge foreign debts. The capital costs for the Philippine reactor are estimated to be nearly four times higher than costs for an oil-fired plant. Moreover, an analysis of energy costs conducted by the Energy Development Board demonstrated that to generate one kilowatt of power would cost $800 for coal, $900 for geothermal, $1,000 for hydroelectric, and $1,500 for nuclear. Significantly, these figures overlook the fact that the Bataan reactor will operate at only 62 percent capacity the first year and will reach only 72 percent capacity after the fifth year of operation.[35] A comparison, made by groups opposed to the nuclear project, of capital and operating costs combined (using 1975-76 dollars) revealed that electricity generated from the nuclear plant is 28 percent more expensive than from an oil-fired plant.[36] Aside from the costs of the plant itself, uranium fuel is both scarce and expensive. Estimates of annual noncommunist world production of uranium are 16,300 tons (United States, 9,000 tons; Canada, 4,700 tons; and South Africa, 2,600 tons), comprising approximately 80 percent of the annual world production. However, some predictions suggest that in twenty years the world will experience a uranium shortage as the energy supply is depleted. The Bataan plant will require an initial load of 600,000 pounds and a yearly reload of 150,000 pounds. At a cost of $40 per pound (1975 prices), and even assuming the unlikelihood of prices remaining stable, the initial load will cost $24 million and the yearly reload $6 million.[37]

The Bataan reactor also has been surrounded with sociopolitical controversy since the project's inception. The residents of Morong were neither consulted nor informed about the project; they learned of the reactor only as their land was expropriated and they were relocated. The regime assumed that there would be little opposition to the plant from the local inhabitants, and at the same time, with martial law in force (from September 1972 to January 1981), government officials saw little need to convince the people of Morong of the plant's necessity.[38] Nevertheless, there have been numerous protests over the project, some resulting in allegations of human rights violations by the government. A typical example of intimidation occurred in an open forum between the NPC and Morong residents, attended by the military and local police. When a Methodist minister asked how the government was going to deal with the problems of pollution, a military officer reportedly responded: "*Siguro aktibista ka, ano? Ipaaaresto kita.*" (Surely you are an activist, are you not? I will have you arrested.)[39] But perhaps the most publicized case was that of Ernesto Nazareno, an antinuclear activist accused of assisting the communist New People's Army (NPA) in organizing plant workers for sabotage and of collecting money for explosives to destroy the facility. Nazareno testified that while under detention for almost three months he was subjected to grueling interrogation and torture, including fist blows to his upper torso and dunking his head into a toilet bowl. Although Nazareno was required to report regularly to the Philippine Constabulary (PC) upon his release, he has been reported missing since June 1978 and is rumored to have been "salvaged" (an euphemism for murder) by the military.[40]

Regime Response to Critics

President Marcos has defended the Westinghouse contract on the following grounds: that the $1.1 billion price tag, despite the Spanish deal with Westinghouse, is lower than similar plants in South Korea and Egypt and that Westinghouse was willing to assume greater risks than General Electric in the eventuality that the plant would not operate.[41] On other issues the NPC and Ebasco have denied the presence of any earthquake faults on or near the reactor site, while Minister of Energy Velasco has defended the project on other grounds, stating: "As to nuclear power, we have included it in our energy development strategy because we have sufficient known reserves of uranium which can keep a nuclear plant going. Safeguards in its operation are fairly well established, and it is said in this regard that it is less likely for radiation from nuclear waste to affect people than it is for a meteorite to kill a person while walking home from work."[42]

Westinghouse also has issued statements on the safety, economy, and reliability of the reactor. Walter Wilgus, a Westinghouse vice president and project manager for the Bataan plant, assured the Philippines that it can recover its investment within ten to fifteen years, pointing out the benefits from foreign exchange savings, and argued that operational costs of the reactor will be lower than those of an oil-fired plant, though admitting the capital costs of the former are higher.[43] In a similar vein, Ian Forbes of the Energy Resource Group, Inc., a consulting firm hired by Westinghouse to assess plant safety, stated that the reactor can withstand severe earthquakes, contrary to assertions by critics, and that the site is the best of five areas considered for the project. Reportedly, "the bedrock on which the plant will rest is superior to any proposed."[44] In testimony before the Puno Commission, David Ferg, the senior Westinghouse nuclear engineer, stressed that the reactor had three safety mechanisms; that the cooling system can absorb heat for twenty to thirty days; that contaminated water will be contained in isolation; and that the building can withstand an earthquake of 7.5 intensity on the Richter Scale. Yet the corporation's assurances were not unconditional, as indicated by an exchange between Puno and Ferg:

Puno: Suppose it [the building] collapses, would Westinghouse pay for the damage wrought?
Ferg: No.
Puno: You are asking us to believe you but you will not pay for the damage if the building collapsed.[45]

Implications of the Philippine Choice

The Philippine decision to purchase a nuclear reactor must be judged, at least in part, in terms of the regime's development goals. According to President Marcos, these are aimed at "the improvement in the well-being of the broad masses," which "means getting down and reaching the poorest segments of our population: the urban and rural poor, the unemployed, the underemployed, the homeless dweller, the out-of-school youth, the landless worker, the Sacada [migrant sugar worker], and the sustenance fisherman."[46] Thus, as a partial measure of how well the nuclear reactor will contribute to the president's development objectives, we will focus on an analysis of the beneficiaries of the power, the distribution of the electricity, the sociopolitical costs/benefits of the project, the international linkages undergirding the nuclear deal, and the reactor's potential contribution to Philippine self-sufficiency.

(1) The greatest demand for increased energy comes from an industrial sector heavily influenced by foreign investors. Commercial entities close to the

reactor that will become major users of energy are the Bataan Export Processing Zone (BEPZ), a free-trade area with access to cheap labor for assembling export products of foreign companies, and numerous industrial complexes in the Manila metropolitan area. Among the heaviest users, for example, will be Marinduque Mining and Industrial Corporation, Procter and Gamble Philippines, and Colgate Palmolive Philippines, all of which have foreign connections. That these companies will be major electricity users is part of a conscious policy of government technocrats to attract foreign investment to help generate economic development similar to the phenomenal growth of Taiwan and South Korea since World War II. The government holds that the foreign corporations provide jobs, access to foreign markets, needed capital, technology, and managerial skills as well as provide better products at lower prices and needed revenues through increased taxes.[47]

Yet critics of the regime's development program argue that foreign investment is hardly an unvarnished success and has failed to bring rapid economic development to the country. Data demonstrate that foreign investment has resulted in a net outflow of income as companies remit profits, after having used local resources for investment, while, concomitantly, many local firms are forced out of business because of competition due to foreign corporations' advantages in economies of scale, in their ability to pay higher wages, and in their relative ease of obtaining finance capital. Moreover, there is very little transfer of technology because research and development takes place in the country of the parent firm. What is brought to the Philippines is packaged technology, requiring only cheap assembly labor; although jobs are created, the positions often have relatively few linkages with the local economy. Finally, the working conditions for local unskilled laborers are characterized by exploitation, redounding to the economic benefit of the foreign enterprises.[48]

(2) The nuclear power plant's electricity will be inequitably distributed within the country. As a corollary to the first point, even if a portion of the nuclear-generated power is diverted for household use, the cost of the electricity still would be beyond the reach of a large segment of the population. Nuclear power, as previously indicated, is an expensive means of generating electricity. Although the government is committed to distributing it equitably, the plant, ironically, will not benefit the great majority who are unable to afford electricity, much less an array of electrical appliances. Clearly, with the top 10 percent of the population accounting for nearly 40 percent of family earnings and the bottom 40 percent less than 20 percent of household income, few have full access to the advantages derived from the use of electricity. The poor are forced to allocate a larger portion of their meager incomes to food consumption. Significantly, agriculture as a sector of the economy uses less than 3 percent of the nation's electrical output. In contrast, 90 percent of the country's power is consumed in urban areas that contain only 30 percent of the population.[49]

Additionally, according to the findings of a World Bank–Asian Development

Bank mission in 1980, noncommercial energy use in the Philippines added up to 25 percent of total consumption, divided between households and industries, and consisted almost entirely of locally available biomass materials.[50] This echoes 1978 statistics of the National Economic and Development Authority (NEDA) where out of 6.2 million households (1.9 million urban, 4.3 million rural), only 2.8 percent (8.3 percent urban, 0.3 percent rural) used electricity for major household activities such as cooking. Nearly 80 percent of the households use wood (49.3 percent urban, 92.5 percent rural), and the rest used kerosene (10.9 percent), gas (5.7 percent), charcoal (0.4 percent), and other fuels (1.0 percent).[51] Yet it is difficult to reconcile these facts with a statement by Imelda Marcos to the National Rural Electrification Cooperative Association (USA) that the regime is "proud of this program [electrification] because it literally lights up our people's lives. It plugs them into the mainstream of national life through the electronic mass media. It multiplies their productivity a thousandfold."[52] Families that must, out of necessity, use biomass materials for cooking are unlikely to possess the means to plug "into the mainstream of national life" for some time to come. In the meantime, electrical use, including power from the nuclear reactor, will continue to be maldistributed to the advantage of the well-off and foreign corporations.

(3) Construction of the plant will not lead to the democratization of social and economic opportunities. Some have argued that the nuclear project represents a nondemocratic trend in terms of the concentration of economic resources, and, as such, is inappropriate because it deprives the nation of funds needed in other more vital areas of development such as land reform and rural transformation. For example, over 30 percent of the country's children under age six suffer from severe malnutrition, and among the leading causes of death are diseases associated with poor water supplies and sanitation. According to the USAID study, the rural poor in the Philippines "do not themselves place a high value on the acquisition of household electricity. When villages without electricity were polled about their preferences, electrification is low down on the list, with the highest priority being given to services like health and water supply."[53]

It is ironic that the same government committed to social justice and economic development for the masses demonstrated such insensitivity to the dislocation of 11 thousand persons living in the vicinity of the nuclear reactor. Plant site construction has led to a loss of grazing and orchard lands, the elimination of rice paddies, the destruction of fish-spawning grounds, as well as the displacement of about one hundred families, most of whom are subsistence farmers and fishermen. Such government actions are not isolated and are repeated in other parts of the country undergoing "development." For example, agribusiness expansion in Bukidnon and Davao del Norte by subsidiaries of foreign multinational companies and large Philippine corporations and the construction of the Kawasaki plant in Misamis Oriental had led to confrontations with local inhabitants, as have development of hydroelectric projects in Augusan and Kalinga-Apayao, with small farmers often losing land and property without just compensation from the government.[54]

(4) The construction of the nuclear power plant reflects a harmony of interests between local elites (center of the Periphery) and foreign elites (center of the Center) and will not lead to the goal of ultimate self-reliance. The decision to build the nuclear reactor was a confluence of interests between Filipino elites, such as government technocrats, local businessmen, and indigenous managers of foreign firms, at the center of the Periphery (Manila) and foreign elites with economic interests both in the center of the Center (the United States in this case) and the center of the Periphery. By supporting the project both elite groups are attempting to maximize competitive advantages, but, as with all neocolonial or dependency relations, the center of the Center elite appears to gain the most. In an immediate sense, Disini and his corporate associates profited from the Westinghouse contracts, and government technocrats can point to a future source of energy for economic development. At the same time local industries dominated by foreign concerns are reassured of a hospitable climate for additional profits. The symbiotic relationship persists, even flourishes, with the regime officially encouraging foreign investment and increasing numbers of local elites identifying with the success of foreign firms, and the government, in turn, pledging to maintain political and social stability to ensure the retention of the foreign corporations.

Such a symbiosis forms a bridgehead over which the harmony of interests is extended between the foreign and local elites, often to the disadvantage of those on the periphery—especially those on the periphery of the Periphery. Within the context of the nuclear power plant, nuclear energy production will sharply reinforce the dependence of the Philippines on the United States, creating a high degree of technological dependency. Moreover, the United States no doubt will be one of Manila's major uranium suppliers, additionally linking the Philippines to the American orbit rather than providing a basis for increased self-reliance.

Conclusion

Any study on technological choices also must stress the importance of the class interests of those making the choices, as well as the set of social, economic, and ideological values by which decisions are made. The choice and use of a particular technology are not politically neutral, as some would have us believe (that is, those who say what is basically involved is choosing the task first and then seeking the most rational way of accomplishing the task).[55] One interpretation regarding the role of technology in any given society is that technology is "designed to support a dominant ideology," i.e., the ideology of industrialization, synonymous with modernization and Westernization.

Industrialization provides an apparent rationality for—and hence appears to legitimate—policies of an exploitive nature. It preaches emancipation through the machine, and has indeed been successful in raising the standard of living for many; yet at the same time it is used to justify the increased domination and oppression of man that has been made possible by the machine. It preaches the social equality and democratization that the machine will bring; yet although it has broken down many traditional class barriers, it is used to legitimate and promote new class divisions and inequalities. Above all, industrialization preaches the political neutrality of technology, representing it as merely a tool to be used for good or ill; yet it produces a technology which is a direct reflection of the ideology of advanced technocratic society, namely the dominance of "scientifically rational thought" and of authoritarian forms of social control over all other interpretations of human experience.[56]

The Philippines, according to President Marcos, is attempting to telescope into a handful of years industrialization, economic and social modernization strategies that took over one hundred years to carry through in already developed countries.[57] Crucial to Marco's industrialization program is the presence of centralized and complex technologies as well as local elites and technocrats supportive of this kind of technology in particular and this growth strategy in general. The choice to employ nuclear technology in generating expensive electricity is an extension of this ideology of industrialization; it reflects the need to sustain particular political interests, economic activities, and a certain pattern of consumption most visible in a small segment of society.

André Gunder Frank has described, in a nutshell, the technological imperialism that occurs between the Center and Periphery nations:

The problem of technology and its diffusion arises out of the same monopoly structure of the economic system on the world, national, and local levels. During the course of the historical development of the capitalist system on these levels, the developed countries have always diffused out to their satellite colonial dependencies the technology whose employment in the colonial and now underdeveloped countries has served the interests of the metropolis; and the metropolis has always suppressed the technology in the now underdeveloped countries which conflicted with the interests of the metropolis and its own development.... Far from diffusing more and more important technology to the underdeveloped countries, the most significant technological trend of our day is the increasing degree to which new technology serves as the basis of the capitalist metropolis' monopoly control over its underdeveloped economic colonies.[58]

The degree of Philippine dependence on the United States, a premier Center nation, can be fully grasped only within the total picture of the society—economy, politics and government, and history. What we have attempted to show is that within this setting a nuclear energy policy has been developed that tends to reinforce Philippine dependency, especially with regard to a very dangerous kind of technological dependency.

7. Intermediate Technology in Newly Industrialized Countries: Two Cases from South Korea

Martin H. Sours

Within the overall context of the interconnection of technology and development, this chapter focuses on two specific cases within a single country, South Korea (more formally known as the Republic of Korea, ROK), and hereinafter referred to simply as Korea. The Korean situation is meaningful in that it is a representative nation-state drawn from the group of newly industrializing countries that are now appearing in the development literature. These states share the general characteristics of having been underdeveloped but are now achieving more rapid growth than other states previously included within the same categorization; thus, they often are singled out for special treatment in the development literature in an effort to detect relevant characteristics of global applicability.

Korea is noteworthy because it shares with other countries struggling through various stages of underdevelopment several general as well as specific disadvantages. Korea has a high population density, a weak natural resource base, high economic interdependence and dependence with the rest of the world, and an initially poor infrastructure. In addition, several characteristics of the Korean situation have been disadvantageous to development from the country's inception after World War II.

Korea was a colony of Japan from 1910 until 1945, during the period of Japanese imperial expansion. As such, it suffered an arrested cultural development. The traditional value system of precolonial Korea was Confucianism, in which a small, conservative elite presided over an agricultural society in which a societal steady state was valued. Japanese colonial rule, while imparting some techniques of modern economic and technological development, did so within an imperial and colonial framework that tended to detract from their desirability and acceptability. Further, World War II left the nation divided into two states, each controlled by mutually hostile and exclusive regimes. The 1950–53 Korean War devastated the South and left the nation divided. The economies of North and South Korea are complementary in that the industrial base and mineral deposits are predominant in the North, while the South has been largely rural and agricultural. The political system that emerged in the South has been characterized as highly centralized and authoritarian, with unstable and irregular patterns of change in the regime. The North has been continuously controlled by an authoritarian Communist party led by Kim Il-sung, who has fostered an extreme form of the cult of personality around both himself and his family.

Given this unpromising setting, this chapter will develop two sections, first, a brief overview of the current economic and political environment within which technological change operates today, and then two case studies will be presented of firms and the intermediate technologies they employ. The original materials for these cases were developed through open-ended interviews and correspondence and are presented for the purpose of discovering some general patterns of success in the development process that do not rely on large-scale projects or international governmental or nongovernmental assistance.

The Current Korean Scene

Korea achieved widespread attention in the mid-1970s as a country on the verge of creating a second "Japanese miracle" and was heralded as one of the most successful of developing societies. As described by the World Bank,

> Korea was one of the poorest developing countries (in 1961) with heavy dependence on agriculture and a weak balance of payments financed almost entirely by foreign grants. By 1976 it had become a semi-industrial, middle-income nation with an increasingly strong external payments position and the prospects of eliminating its current account deficit in the late 1970s. GNP grew over the period at the average rate of more than 10 percent a year, and per capita income tripled in real terms.[1]

The elements that contributed to this rapid turnaround were both internal to Korea and externally rooted in international affairs. Internally, the government of Park Chung-hee, which came to power in the early 1960s, established economic growth as a national priority and directed resources toward that objective. Within the context of a market economy, the government guided the investment process through low-interest loans to favored enterprises, while mass consumption was retarded. Thus, selected, highly leveraged Korean industries could undergo expansion in a controlled and favorable domestic environment.

A major stimulus for this expansion came from the international situation. The constant and real threat from North Korea allowed the Park government to operate with authoritarian powers, thus negating virtually all forms of dissent. Concurrently, the normalization of relations between Korea and Japan in 1965 established the framework for Japanese commercial penetration of the country so that, in fact, Korean firms could become labor-intensive subcontractors for Japanese enterprises at the point when labor-intensive production was being phased out in Japan. More importantly, additional external foreign exchange was earned in the late 1960s and early 1970s by Korean combat troops introduced into the Vietnamese war and by Korean construction companies that successfully bid on and received contracts for construction projects in the Middle East. Because currency controls were absolute in Korea (as a national security measure), massive net amounts of capital flowed into Korea from these projects, and these funds, in turn, were used for reinvestment and debt service under government direction and guidance.

Two other things contributed to Korean economic expansion. At a time of rising Japanese affluence, Korea became a popular destination for Japanese tourists. With Korean labor costs maintained at low levels by governmental policy, and thus unions, strikes, and other market forces that would ordinarily raise such costs restricted, tourism, a labor-intensive industry, was very competitive with other alternatives available to Japanese tourists.

The same cost and cultural factors contributed to low-cost, labor-intensive manufacturing. Without increases in labor costs, price-sensitive goods could be produced in Korea for world markets. Thus, the export-led economic growth strategy pursued by Korea during the 1970s proved highly successful, and even though debt from foreign borrowing continued to mount, the credit rating of the country remained excellent in international financial circles, creating a debt-financed, high-growth economy that fed upon itself.

By the end of the decade, however, the worldwide twin problems of high inflation and recession put severe strains on the Korean economy. According to U.S. Embassy figures, economic statistics for 1980 revealed a bleak picture, especially for an economy that had shown uninterrupted growth for nearly two decades. Oil price increases and a slowdown in the world economy, which determined the market demand for Korean exports, as well as wage increases necessitated by domestic inflationary pressures created severe economic problems. Import prices, especially oil for which Korea is totally dependent, have risen, while export price increases have been constrained by competition from Korea's Asian export rivals. To spur export, the Korean government ordered a 20 percent devaluation of the Korean currency, the won, in January 1980. Additionally, the won had been pegged to the U.S. dollar so that the won floated downward beyond the formal devaluation rate, creating a total devaluation effect of about 36 percent in 1980. This created inflationary pressures in the country by driving up the cost of raw material imports.[2]

These problems were exacerbated by political instability resulting from the assassination of President Park Chung-hee in October 1979. The entire year 1980 was consumed by the political shifts that followed. Initially, a period of liberalization appeared to unfold in which movement toward a type of controlled participative government seemed possible. However, a coup on December 12, 1979, referred to informally as the 12/12 incident, brought into real power a group of junior generals, led by Chun Doo Hwan. Most disturbing from an international development perspective were the "Six Points" issued by the new military leaders. They included a rejection of the primacy of private interests that might conflict with the general public good—a view that could be interpreted as being negative toward entrepreneurship—and a cautionary note with regard to excessive individualism.[3]

Subsequent events appeared to cloud the general environment of the country. Establishment politicians were purged, riots in the southern city of Gwangju

(Kwangju) were repressed by military force, the leading opposition figure, Kim Dae Jung, was arrested, tried, and sentenced to death, and the constitutional president, Choi Kyu Hah, was in effect forced to resign, paving the way for the formal ascension to power of General Chun. A new constitution issued in late 1980 provided the framework for Chun's formal election as president.

All of these events cast a long shadow over the country, which was apparently lifted by the Reagan-Chun summit meeting in Washington, D.C., in early February 1981. Almost at the same time the death sentence imposed on Kim Dae Jung was commuted to life imprisonment, coupled with other signals that more normal conditions were returning to South Korea. Thus, while no proof of linkage has been publicly offered, the appearance has been created that the Reagan administration has executed a policy shift from confrontation to support of the Chun government, with the tacit counterpolicy in Korea of moderation, stability, and a return to normalcy, especially with respect to economic matters.

The disturbing political and economic events of recent years suggest that Korean economic growth and political stability are endangered. While this may be true at the macro level, the cases examined below indicate that in selected situations micro-economic factors may simulate entrepreneurial success or failure quite independently of overall macro trends in the society and economy. This aspect of economic success involving appropriate technology is clearly demonstrated in the case of the most successful Korean company now producing and distributing "pipehouses."

The Abundant Harvest Pipehouse Company

Originally, farmers in Korea copied the Japanese system of constructing simple greenhouses out of bamboo and covering them with vinyl. The Japanese farming community, supported by the Japanese government and the postwar land reform program, formed cooperatives and progressively refined the system for constructing and maintaining these greenhouses. After the normalization of relations between Japan and South Korea in 1965, Korean farmers became aware of these procedures, but the Korean agricultural community was about ten years behind that of Japan. In particular, the bamboo and vinyl greenhouses were weak and short-lived in the comparatively harsh Korean climate. Thus, they were constructed with a life expectancy of only one year, or one growing season. Many greenhouses suffered from breakage and disintegration. There was an objective need for an improved system in Korea, particularly as this need related to overall shifts in Korean agricultural production.

After World War II the southern part of the Korean peninsula was largely agricultural. But after the devastation of the Korean War, the industrialization program created a shift in the nation's economy in which an increasingly smaller percentage of the population and Gross National Product was derived from the agricultural sector. This shift required increased productivity from Korean agriculture, and, concurrently, the food consumption patterns of the country became more diversified. The rise in demand for fruits and vegetables, if it was to be met by domestic production, required more sophisticated agricultural techniques. Because domestic self-sufficiency was a national goal to preserve foreign exchange for industrialization, conditions were favorable for the inauguration of new agricultural techniques.

In this environment a Korean entrepreneur went to Japan in the early 1970s and studied the development of greenhouses there. This trip in itself required some ingenuity and assistance, since foreign travel has not been a natural right for Korean citizens. However, a Japanese associate from previous ventures assisted by inviting the Korean businessman to Japan, thus fulfilling the Korean government mandate that foreign travel be by invitation and of no cost to the Korean citizen to prevent any loss of foreign exchange (these restrictions are somewhat less stringent today). Once armed with a Korean passport and Japanese visa, the Korean businessman observed the scene in Japan, with particular reference to a new type of pipe joint. These pipes, joined together, could become the frame for a new kind of greenhouse modeled after the original bamboo greenhouses but with distinct advantages in longevity and durability.

Upon his return to Korea, the businessman began the long process of patent application to the government. This required the construction of sample pipe joints that had to be stored. There were numerous delays, resulting in loss of samples, financial problems with bankers, lack of venture capital, and, because the businessman was not politically connected, extra delays in the processing of his application. These delays were finally overcome in 1976 when the patent rights were granted.

Since the businessman's personal financial resources were literally exhausted and his family's financial position had deteriorated, start-up venture capital was then critical and not available through personal channels. However, the businessman had a friend and associate from previous ventures in Japan who was a successful Korean resident there. He was looking for an investment opportunity. He loaned the start-up capital to begin the venture, and thus the high interest rates that normally would be charged by Korean financial institutions were avoided. Ultimately, with the success of the venture, this loan was repaid in full; now the business is privately held and completely debt-free.

The specific technology involved was a set of pipe joints used to join various sizes of pipe together to form the frame for the newer and stronger greenhouses. The market demand for such systems came from the increased market for vegetables, particularly for the higher grade of vegetables that could be grown faster under controlled greenhouse conditions. The basic marketing issues at the start of production and distribution were twofold: where to begin sale and distribution, and which farmers could afford the system.

Initially, the product was displayed and sold in the southern part of Korea around the port city of Pusan. While the more northern climate closer to the capital city of Seoul might appear more logical, the selection of Pusan rested on sound knowledge of the local market situation. The farmers in the south were already in the vegetable business because of the milder climate, and Pusan's relatively close proximity to Japan made the introduction of new Japanese technology easier to accept. As a result, initial sales were made to growers of such traditional Korean vegetables as spinach, cucumbers, Chinese cabbage, and radishes.

Initial sales were made to relatively well-established farmers who could afford to switch from bamboo to pipehouse. If a farmer owned his land, he could obtain a low-interest loan from government agricultural banks for from 15 to 30 percent of the cost of the pipehouse. Thus, relatively richer farmers formed the demonstration group for sale of the system, and it spread north from the Pusan area to other agricultural sections. Since this product was a new one, providing a technological jump from bamboo to pipe, its market potential consisted of all farming households in the country, or about two million units. To date, some 30 to 40 percent of these households have converted to pipehouse, or about 800 thousand units, of which the Abundant Harvest Company, being the first in this field with an established reputation, has the majority.

Being first was important, for the enforcement of patent rights in Korea is relatively weak, so once the system was demonstrated as successful there were companies ready to act as imitators. However, within the Korean cultural context the notion of established reliability is highly influential. Thus, it became a status symbol not only to have a pipehouse, but to have the best one produced by the company that introduced the product. Therefore, the maintenance of quality control became crucial.

Here again the entrepreneurial skills of the founder were of central importance. The pipehouses may be constructed out of three sizes of pipe, 19 mm, 22 mm, and 25 mm. The company established its own fabrication plant in Seoul and contracted with the semipublic Unified Steel Corporation for supplies of steel pipe. The pipehouse company employs skilled technicians who are graduates of technical high schools with many years experience in the steel industry and who work full-time fabricating and installing each unit on a custom-design basis. The zinc-coated steel pipe employed resists rust. This rust-resistant feature is necessary because of the large amount of moisture generated inside the covered pipehouses. The constructed pipehouses are covered with heavy-duty tent canvas, which is more durable than the original vinyl.

The average pipehouse covers a 100 *pyung* area, about 1,800 square feet. The system became so attractive that farmers' cooperatives in Korea became customers, creating a proliferation of secondary uses for the pipehouse. Depending on their size, pipehouses also were employed as hothouses for flower production, sheds for goats, pigs, chickens, and cows, and as inexpensive storage sheds on the roofs of urban dwellings. There was even an attempt to use them as modest leisure houses at vacation sites in support of the tourist industry.

To maintain its large share of the market, the company maintains its central office and factory in Seoul and advertises extensively in national daily newspapers and farmers' journals. There are no fixed prices for units. Rather, prices are determined on the basis of design requirements, with additional charges assessed for overtime production, rapid delivery and setup, and transportation costs. This flexibility in pricing is necessary to adjust for the rapid inflationary pressures that presently operate on the Korean economy. Thus, the company is able to prosper at a time when some large Korean firms are experiencing cost-squeeze pressures and resultant financial difficulties.

The next case examined is derived from the more internationally recognized electronics industry, the Sunil Electronics Company. Here we have a situation in which entrepreneurial skills and appropriate technology interact in the highly competitive export market.

The Sunil Electronics Company, Ltd.

During the 1960s a consumer electronics industry began to emerge in Japan. Its products included radios, tape recorders, and components for stereo systems. As Japanese firms grew in size and sales, they began to use the assembly method of production, employing lines staffed with female workers in their late teens. This growth kept pace with demand until the late 1960s when demand accelerated rapidly along with the general Japanese economy. This produced a rise in the standard of living in Japan, which in turn pushed up wage rates. Because a major element in production costs consisted of labor wages, the need for alternate sources of labor grew in Japan at the same time as Japanese population growth stabilized. This created a shortfall in the supply of labor. About the same time there occurred a general shift of Japanese workers away from the relatively low pay and low prestige of assembly work into skilled work and the service sector.

After the normalization of Korean-Japanese diplomatic relations in 1965, Japanese businessmen came to Korea to explore the possibilities of establishing factories to produce electronic components in Korea based on Japanese models. The Japanese brought with them the knowledge of electronic manufacturing at precisely the time when the Korean government wanted to embark on its industrialization program. Thus, the Japanese firms offered not only technology, but also a market for the output of Korean producers who entered into business relationships with Japanese firms. Japan therefore became the market and also the source of raw materials. These materials consisted of special metals and wire produced to exacting specifications. Additionally, venture capital was provided by the Japanese concerns. Therefore, the Korean producers became Japanese subsidiaries, in fact, while retaining the appearance of independent firms.

By 1970 the perception of this arrangement in Korea was one of exploitation. The Korean firms sought to divest themselves of contractual links with Japan while retaining the business relationships. In particular, Japanese imports of raw materials were seen as too expensive, leading to high profits for Japanese producers. The Korean firms moved toward copying the entire Japanese production cycle, and with the growth of Korean electronics firms, new customers emerged for electronics components. Concurrently, the increased costs of doing business in Korea led the major Japanese firms to move farther afield to set up assembly operations elsewhere, principally in Southeast Asia.

The experience of Sunil should be seen within this context. It was established in 1970 as a Korean/Japanese joint venture to produce Mylar condensers. At the time of inception, the firm employed three hundred line workers and produced two million units per month. By 1976 the Korean owner bought out his Japanese partners and a completely Korean firm was established. As productivity rose, the number of line assemblers was reduced to two hundred workers, with a supervisory staff of ten technicians and ten managers. With productivity increases, output rose to three million units per month, with half sold to Japanese firms and half delivered to the emerging Korean electronics firms, such as "Gold Star."

The major quantum expansion in technology and systems production occurred in 1979 when ten automatic wire-winding machines were purchased by the firm. These were manufactured in Japan, but they greatly increased productivity since the two products produced by the firm, Mylar condensers (Polyester film capacitor) and Styrol condensers (Polystyrene film capacitor), have as their major production input the winding of wire according to specification. With the automatic winding machines, 50 percent of production could be automated.

In 1980 an additional ten automatic winding machines were purchased, and this time the machines were manufactured in Korea. Therefore, all production became automated, resulting in production runs of six million units per month. The work force also was reduced to only eighty line operators, with the ten technical specialists and ten managers remaining in place. Also a major shift in the market occurred in which five million out of the monthly run of six million units were sold to Korean users, with the remaining one million units still going for export to Japan.

Sunil represents one of about twenty small electronics firms that operate in the middle of the production system of consumer electronics. The average production run for these twenty firms is from two million to twelve million units per month. The salary for the remaining production workers, girls from about seventeen to twenty years of age who have graduated from elementary or junior high school, is $90 per month. The technical supervisors, who are technical high school graduates ranging from twenty-two to twenty-eight years old, earn about $150 per month. The salary range for management is from $300 to $650 per month. These wages are lower than those earned by employees of the large general corporations of Korea, but the wage scales fit into the general wage structure of the country.

With 30 percent of the sale price of the capacitors going for profit, the firm appears extremely successful. But the capital for operating costs and modernization is generated by current sales. The structure of the Korean capital market is weak, and as indicated in the Abundant Harvest case, much start-up capital comes from unofficial sources, in these two cases, Japan. Further, the firm had to purchase outright all twenty of its automatic winding machines, but the firm's customers definitely preferred the automatically produced capacitors because they were standard, conformed to specifications, and were free of human error. Some external financing has been available because the Sunil Company is exporting part of its production. Therefore, it is eligible for special loans, secured by the letters of credit of its overseas customers, at the preferential rate of 12.8 percent interest, well below the market rate in Korea. This money is known informally as export-supporting loans, and the rate is encouraged and supported by the Korean government as a matter of public policy.

Today the company is moving in a capital-intensive direction to cut down on labor costs; in other words, it is following the Japanese model. Workers take three months to train, during which time raw materials used for practice runs must be expended and discarded. With the rise in Korean living standards, competent workers are hard to find who are willing to work in factories, and machines reduce production costs by as much as 50 percent, making them a highly desirable investment. Thus, the firm, while charging a price with a large profit margin or makeup, is reinvesting its profits in new equipment to stay competitive and maintain the capability of supplying both domestic and international customers according to their ever-higher specification standards.

Conclusions

Both of the companies examined here have developed an institutionalized presentation and the appearance of long-term stability. This is important in a country such as Korea where many highly leveraged firms often are forced into bankruptcy because of their weak financial positions. These businesses have established themselves as growing medium- or small-scale firms (the U.S. Small Business Administration classification of fifty employees would put both of them close to the small business category). While one firm demonstrates some aspects of domestic entrepreneurship and the other is a product of an international growth industry, they share certain characteristics.

First, both firms benefited by the spillover effect of industrialization and economic growth in Japan. This suggests some definite benefits both in terms of demonstration and availability of technology that can be exported. The exportation process from Japan may be traced to both "pull" factors, such as the demand for better greenhouses, and "push" factors, such as cost differentials.

Second, there had to be some sort of cultural compatibility in order for the businessmen concerned to communicate. This process may be unique to Korea where there is a public suspicion of Japan because of the Japanese colonial legacy from 1910 to 1945, yet the cultural roots and languages are structurally similar, thus operational communications are greatly facilitated. Japan therefore could act as a laboratory for the development of systems that were relevant to the Korean situation.

Third, the cases do not demonstrate any structured formula for the introduction of entrepreneurship and technology into a society. In the cases examined, the entrepreneurial spirit came from the individual who founded the Abundant Harvest greenhouse company and from the Japanese corporation that introduced the appropriate technology and then facilitated and guaranteed the initial markets for the Sunil Company.

Perhaps most important is the existence within society of a "critical mass" of individuals committed to change and entrepreneurship. Here the Korean situation can generate useful insights. While general economic and political conditions deteriorated badly in 1980, opportunities for the economic growth of selected firms remained. As was shown, these two firms were able to grow in spite of the generally poor economic environment. This growth occurred because of intrinsic demand for their products, good product modification, and their responsiveness to ongoing market conditions.

These examples suggest that research in the area of development and appropriate technology needs to focus on the sources of technology, its precise relevance to microeconomic conditions within industries or segments of an economy, and the effects introduced by private firms in the development of new systems in the intermediate range. As more data are gathered, patterns may emerge from examples such as these that could have general applicability. At present there is a tendency to dismiss the Korean experience as generally inapplicable precisely because of the country's proximity to Japan and the overwhelming Japanese component in the Korean development process. Inquiry along those lines might well lead back to the ongoing academic debate concerning the uniqueness of Japanese modernization and the role of imported and indigenous technology in that process.

8. Weapons Production in Less Developed Countries: A Possibility for Integrating Technology with Development?

Daniel J. Bohlin

> Ataturk turned the problem around . . . by making the modernization of the army dependent on the transformation of Turkish society.
> Henry Bienen, *The Military and Modernization*

In her summary of the literature concerning the effects armed forces have on the development process in less developed countries, Mary Kaldor concludes, "existing research, while representing a beginning, is quite inadequate."[1] This is because researchers usually have treated the military as a side issue not really central to the development process. Yet, after observing the number of so-called "militarized" governments that have emerged in the LDCs, one must question placing the armed forces of these nations to the "side" when evaluating their political, social, and economic development.[2]

Writing in 1980, Richard Falk graphically portrayed the militarization phenomenon that has occurred since 1960 in many LDCs. His work shows the emerging militarism manifested by governments of LDCs in Latin America, parts of Africa, most of the Middle East, and a large portion of Southeast Asia. Falk used many indicators, ranging from "having a military ruler and meaningless elections" to "military used for internal security in non-emergency situations." Noteworthy as one indicator of militarism was a "high percentage of a nation's budget used for military purposes."[3]

When a government uses a substantial portion of its budget for military purposes, it will use some of this money to obtain military weapons.[4] It usually will purchase the weapons either from a foreign producer, referred to here as an arms transfer (AT), or from a source manufacturing the weapons domestically, herein referred to as domestic arms production (DAP). Whether purchasing weapons via the AT or DAP process, or both, a LDC is interacting with a technology—the military weapon systems technology.

This weapons technology transfer is part of the militarization phenomenon that can affect both a LDC's armed forces and its development pattern. I will focus on AT and DAP as aspects of militarization and show how they relate to the development efforts of LDCs. Since it is not within the scope of this chapter to address all the possible interactions between AT and DAP, this presentation will emphasize the negative aspects of AT and the positive potential of DAP. It also will present a plan for eliminating the former while enhancing the latter.

Arms Transfers—Trends and Effects

In comparison to the world's more developed nations, the arming of LDCs is increasing at a rapid rate. The International Peace Research Association (IPRA) has stated that while approximately half of the arms transferred on world markets during the 1960s were traded between industrialized countries, that figure dropped to a mere 15 percent in the 1970s, meaning the LDCs were absorbing 85 percent of the world's weapons transfers during the last decade.[5] Likewise, through the mid-1970s Third World expenditures for arms, in terms of annual rates of change, increased by 10 percent, far surpassing the world's total annual rate of change of approximately 3 percent. Viewed in still another fashion, as a percentage of a nation's GNP spent on weapons, the world average declined from about 7 percent to a little less than 6 percent, while the LDCs' figure rose from about 4 percent to more than 5 percent in the same period.

The implications of these trends for the future are, of course, not certain. The figures show only what has recently occurred; future military expenditures by the LDCs may diminish. Further, one cannot definitively state this trend has slowed Third World development. In fact, for some LDCs this trend could actually have aided their development, since a few Third World countries—Israel and Brazil, for example—have developed lucrative export markets through resale or production of arms, or both. However, given these recent trends and the present levels of tension in the world, one can argue that sizable expenditures on arms by Third World nations probably will continue and may soon accelerate.

Regardless of the trend—in other words, until swords are beaten into plowshares—military weapons or the military technology question will be with us. The arming of a nation-state, developed or underdeveloped, has been viewed traditionally as rising from the need of that nation-state's government to protect its sovereignty over society. In other words, the government must protect both itself and the society over which it has jurisdiction from external and internal threats—real and perceived. Thus, military weapons, regarded as the instruments that guarantee a nation-state's sovereignty, become highly important for most governments.

Looking at LDCs, this need for protection may be more important, in some cases, than the development of their societies. This may be particularly true for nation-states, like Israel, that confront a real, overt external threat. Or, it may be the case of a perceived external menace, which seems to be the impetus behind Saudi Arabia's arming efforts. Likewise, a serious internal problem—Brazil is an example—may produce emphasis on weapons expenditures over some aspects of civilian development. Finally, it can be both a serious internal and external threat (South Africa) that spurs arming and diminishes development possibilities. Certainly there are other reasons; but apparently conclusive during the past decade is the following: Military weapons expenditures, in particular, and the overall militarization phenomenon, in general, were on the upswing for LDCs, while their opportunities for development—looking across the social, political, and economic spectra—were decreasing.

In discussing the militarization phenomenon of LDCs and linking it to their weapons expenditures, Jan Oberg maintains there are three ways for a society to acquire a military capability: buying arms via arms transfers, domestic arms production, or a combination of the two, AT-DAP.[7] To these I would add a corollary to a LDC attaining a military capability. With an arms exporting (AE) capability, a nation adds to its GNP while increasing the demand for its military products, thereby enhancing its own military capability because its domestic production expands. More money goes into research and development for new arms production, and the military-industrial infrastructure expands. Developed nations long have recognized the importance of expanding their arms exports, both from a military capability and an economic development point of view. Thus, the most desirable situation that permits a country, developed or undeveloped, to maintain and enhance its military capability seems to be the DAP-AE combination. With this combination, a noteworthy phenomenon also occurs—a country all but eliminates AT from its military capability program.

Looking at the vast majority of LDCs, this optimal situation is far from reality. On the contrary, most are quite heavily dependent on the world's major arms producers (the United States, the USSR, France, and Great Britain) for the development and maintenance of their military capability. Only India, Israel, Brazil, and South Africa have, in fact, exhibited impressive DAP results, let alone any significant AE.[8] At present AT is the primary means by which most LDCs sustain their military capability. Some LDCs have absolutely no choice, since no means of arms production exist. For others, if the means of production do exist, it may still be cheaper for that country to import weapons rather than to produce them locally. In either case, observers often argue that the impact of ATs on the Third World's development process is quite negative. The following reasons seem to summarize why AT has such an inimical effect on the LDCs' development process.

First, AT increases rather than decreases the dependency of LDCs upon the weapons-exporting nations with which they trade other nonmilitary products. Just buying the arms, especially if done regularly, implies dependency upon the supplier. Purchasing the weapons, however, is only the tip of the iceberg. For example, in addition to the weapons the LDCs must procure spare parts, develop maintenance facilities, and go to the supplier's country for training courses. Generally, the more complex the weapons (that is, the more complex the technology) the more this dependency occurs. Further, except for some oil-producing states that may pay with cash, LDCs usually purchase their weapons through some financing deal with the supplying nation. Thus, they also become financially dependent upon the supplier. Finally, suppliers usually attach conditions regarding a weapon's use or its resale to a third party. In this case, the supplier obtains diplomatic leverage over a LDC and may make substantial policy dictations to it.

Second, the LDCs will be in a technology transfer interaction that may not be to their benefit. The purchased arms—produced by another culture whose weapons technology reflects its value system, plus its military doctrine, strategy, and tactics, as well as other cultural biases—transfer that technology to the LDC. In many cases the weapons, which ideally should reflect the LDC's cultural values, military doctrine, etc., may not be those most suited to the purchaser's particular needs when they come from a foreign source.

Additionally, with technological breakthroughs in military hardware occurring in producer countries at practically an exponential rate, technological obsolescence arrives sooner. The client must, therefore, buy at a faster rate, or risk falling behind. Further, with production costs zooming, economies of scale force producers to sell state-of-the-art equipment to the richer LDCs that can afford it. Also, with the arms export business becoming more lucrative and thus more competitive between established and emerging suppliers, a producer that may not even want to sell a certain product to some LDCs is still under pressure to do so. One sees, therefore, how an underdeveloped country that may have no real need for a particular military technology may, nonetheless, buy it and then have to live with the consequences of the purchase.

Living with the consequences denotes the third and perhaps most important reason why AT does not seem to be in the best interests of LDCs. This is the effect, directed more specifically to the situation inside a client country, that AT may have on a recipient's political, social, and economic development. Concentrating on the economic angle, the IPRA cites several deleterious effects on development:

> (1) Domestically generated economic surplus is diverted from mass distribution and [private] consumption.
> (2) Foreign exchange is diverted from the import of basic consumer necessities and equipment for production.
> (3) External debt burdens tend to increase, except for a small number of petroleum producing countries. Refinancing is customarily associated with the requirement to adopt austerity policies, which further reduce mass consumption and production accumulation.
> (4) Inflationary tendencies are stimulated, and these operate to depress the real earnings of wage labor.[9]

One may deduce other detrimental effects related to social and political development. A worsening economic condition delays improvements in the social services area. Moreover, as a political environment conducive to development actually deteriorates, this may lead to a confrontation between a LDC's armed forces and the rest of society. AT appears, therefore, to be a no-win situation for at least some LDCs and to a certain degree for all of them.

This does not imply, however, that matters related to the armed forces, defense spending, or even an increasing military budget vis-à-vis economic growth are contrary to a LDC's development efforts. In different parts of the world, depending in some respects on the orientation of the armed forces, the military may promote certain short-term interests of the population. IPRA's evaluations have shown that in some instances where the military pursues a "national or state capitalist policy" it can foster some improvement in the satisfaction of basic needs for large sectors of the population. IPRA has observed some LDCs' armed forces promoting improvements in basic needs in one or more of the following ways:

(1) Agrarian reform.
(2) Expansion of employment.
(3) Retention of greater surplus for investment as a result of the expropriation of major industrial, financial, and commercial enterprises.
(4) Rapid extension of educational opportunities and health facilities.
(5) Constraining inflationary tendencies for mass necessities; and limiting the growth of regressive taxation.
(6) The use of units of armed forces for social and production investment.[10]

In sum, considering that (1) Third World expenditures for military weapons are increasing, (2) AT seems quite inimical to the LDCs' development needs, and (3) a LDC's armed forces can in certain circumstances apparently promote development, do some possibilities exist for fostering development or at least diminishing the hazards to it?

Domestic Arms Production—A Deeper Look

Writing around 1975, a collection of social scientists known as the Hamburg Group determined that there was a considerable lack of analysis concerning DAP and intermediate, military-type technology being used to develop the Third World's armed forces and, as a consequence, aiding the overall development process. The group cited two areas as being wholly neglected. One area was "The development of adapted arms technologies particularly designed for the market of peripheral countries (low price technology, intermediate technology)." A second area requiring research was the study of "the consequences of military transfers in conditioning the character of technology transfer thus determining the feasible strategies of development and industrialization as well."[11]

Noting that only four LDCs were engaged in DAP in 1945 (Argentina, Brazil, South Africa, and India) while there were about forty-six in 1976, a propitious moment apparently has arrived for development-related research on DAP. Some researchers already have discussed different aspects of the DAP question.[12] In 1974 Dieter Ernst, a critic of some transfer of technology approaches as applied to development, noted there were attempts being made "to induce the growth of local industries via the establishment of local arms industries."[13] For example, pursuing a DAP-type policy can foster growth of tool and machine factories as well as industries transforming raw materials. This is the "spin-off" concept whereby introduction of some technology indirectly, as much or more than directly, adds impetus to economic growth and development. Ernst claimed though that this would "fulfill nearly all the preconditions to increase significantly the technological dependence of these countries."[14] Picking up the theme, other critics of DAP policies simply maintain there is no way for the LDCs to compete technologically against industrialized nations and come out on top.[15]

But, having seen that the industrialized nations' more advanced technology may not be the most appropriate for another culture, this particular critique has a serious weakness. Appropriate technology, as opposed to a more complex technology frequently found in industrialized countries' weapons programs, is often the key to the armed forces' successes in many of their endeavors. If a weapon's complexity overwhelms the user, if it cannot be properly maintained, or if it does not function in some operating conditions, it is not appropriate. When AT is involved, the technology transferred is almost always designed for the producer's armed forces, not the LDCs' military counterpart. This further increases the possibility of the technology being inappropriate, thereby adding to the argument for indigenously produced arms.

Knowing that DAP most likely will engender elements of appropriate technology that will conform to the real needs of a LDC's armed forces makes DAP attractive, and not a liability, as its critics want us to believe. Further, DAP becomes even more intriguing when one considers that it also could be designed to aid a LDC's development process, something most foreign military technology does not attempt to do. Before constructing a possible framework for DAP that could be appropriate for an underdeveloped country's armed forces and its development program, let us briefly examine some different analytical approaches that may give clues for a development-oriented model.

Background for a DAP-Development Model

There seem to be two major classifications of research groups providing insight into making DAP and development compatible. Neither completely develops the direct linkage between DAP and a LDC's development process, but each does give some guidance that can contribute to the creation of a framework.

Oberg labels this first group "defense economy research," while Kaldor calls it the "military in the allocation of resources" approach.[16] Works like Emile Benoit's *Defense and Economic Growth in Developing Countries* and Gavin Kennedy's *The Military in the Third World* make up this category. Benoit deals specifically with the interaction between an increase in defense spending and its effects on economic growth. Kennedy approaches the same topic more subjectively, along with other aspects of the military in LDCs. Benoit found in his empirical study of data gathered on India, both before and after its war with China in 1962, that there was a positive correlation between the burden of defense spending (the share of the Gross Domestic Product, GDP, devoted to defense) and the rate of nonmilitary output, or civilian GDP.[17] However, since high foreign aid made available to India after 1962 correlated with the GDP growth rate, the association between a defense burden increase and civilian GDP growth may be spurious.

Hypothesizing different possibilities for India's defense spending rate—increasing, staying the same, or decreasing—Benoit then projects different economic potentials. Noteworthy is that Benoit talks only about development in terms of economic growth (increasing GDP) and apparently assumes that the so-called trickle down effect will pass this growth on to other groups in society, which, in turn, will enhance social and political development.

Both Kennedy and Benoit seem influenced by relatively optimistic American writings that regarded the military as a key factor for spending development.[18] In their view, the armed forces contribute to manpower training, education, and development of skills that can be used in civilian endeavors upon demobilization. The arguments are plausible; yet they assume a value orientation that is problematic.

The second analytic approach, called "Armament-imperialism" research by Oberg, manifests a Marxist orientation.[19] Arms and technology transfers are seen as imperialistic. Researchers in this category, like the Hamburg Group, argue that the militarization of society impedes development, which should be oriented toward meeting basic needs. In their view, the world has embarked on a permanent, ever-intensifying arms race that will lead to stronger, more militarized states needed to repress growing worldwide social contradictions.

While critical of mounting Third World militarism, these researchers have yet to produce a development model that could reverse this trend. However, an emerging approach based upon the idea of a New International Economic Order (NIEO) advocates a nondependent type of development linking the LDCs and the more prosperous, industrialized nations.[20]

The nascent DAP capacity of several newly industrialized countries is an instance of self-reliance by parts of the Third World. For various reasons, all seemingly related to national security interpreted in its broadest sense, some NICs have exhibited high motivation to minimize their dependency on foreign arms producers. Countries like India, Israel, South Africa, Brazil, and Argentina have demonstrated an increasing capacity to become domestic arms producers.

In addition, two of these countries, Israel and Brazil, have developed quite an export capability for some of their armaments. Israel supplied arms to Argentina during the Falkland Islands crisis and sells its Uzi machine gun to clients all over the world.[21] Brazil has made substantial progress in the armored car business, especially after it successfully negotiated a sale with Colonel Kaddafi's Libyan army. One reason for the purchase from Brazil was that the Brazilian vehicles were easily maintained by Libya's young and inexperienced army.[22] Contrast this to evidence showing the need for appropriate technology vis-à-vis the armed forces with which it interacts. During the Indo-Pakistani War in 1965, "Where the education and skills of the soldier were roughly matched, sophisticated military equipment actually proved a handicap . . . Pakistani soldiers were unable to operate the automatic control needed to fire the guns of their Patton tanks, and this considerably hampered their offensive."[23]

Still other LDCs, unable to produce independently the weapons they "need," have made a collective attempt to dissociate from their traditional arms suppliers. Using capital surplus, four nations—Egypt, Saudi Arabia, Qatar, and the United Arab Emirates—pooled $1.4 billion in 1974 and forged a collective self-reliance weapons production agreement.[24] Billed the Arab Organization of Industrialization (AOI) and "using a foundation of Middle East oil money, Western technical and managerial aid and Egyptian military and production experience extending from hand grenades to supersonic aircraft," the Arab consortium planned to produce and to export.[25] However, when Egypt signed the peace treaty with Israel in 1979, the AOI dissolved.

A Framework for Domestic Arms Production

Having shown the need for more emphasis on DAP versus AT and also noting the emerging trends, a DAP strategy should construct itself around these observations. Maintaining the dual goals of preserving the sovereignty of the nation-state and fostering social, political, and economic development in the Third World, the following general framework seems a plausible plan for domestic arms production in LDCs:

(1) Based on the NIEO approach to nation-state sovereignty and a benign dissociation philosophy, LDCs should adhere to a doctrine of self-reliance or collective self-reliance when arming their military forces.

(2) Therefore, LDCs must minimize arms transfers. Should arms transfer become necessary, Third World countries should transfer the arms among themselves, preferably among nations within the same regional system.

(3) In place of arms transfers, LDCs must emphasize domestic arms production that uses appropriate technology according to the following sequence: (*a*) Production under license with appropriate technology obtained first, whenever possible, from other regional actors. (*b*) License modifications occur in order to adapt the weapons to developing needs of the consuming nation. (*c*) As obsolescence occurs, producers must conduct their own research and development, which should center first on items suitable for the producer's needs, then for export.

(4) Researchers must evaluate any introduced weapons technology for the effect it will have on the LDC's armed forces, their relationship to that country's society, and the LDC's overall development process. If, according to some predefined criteria, the technology contributes constructively to these areas, the

LDC should accept it. This positive contribution, in essence, defines the appropriateness of a weapon's technology for a LDC.

This broad framework is adaptable to the various stages of development and regional interaction presently found among LDCs throughout the world. Mindful, though, of the aforementioned trends, adoption of this general DAP framework should start with those nations, like the NICs, that already have started DAP and in some instances, DAP-AE. The NICs must continue to further develop their DAP-AE potential while trimming their own arms imports needs. In addition, donating or selling at low prices surplus weapons to poorer LDCs in a NIC's region is an encouraging aspect of this approach. Essentially, some of these NICs could become the regional leaders around which peripheral LDCs will gravitate.

Regarding the more financially prosperous LDCs, like those with capital surplus but which still are not economically developed, a collective self-reliance orientation is more feasible. Excess capital that probably would never be used in the development process (money in foreign banks, luxury-goods purchases instead of capital investments, etc.) could be used to start organizations like the defunct AOI. Given time, this approach could be useful for diminishing intraregional rivalries in some areas of the world where increasing arms transfers probably have contributed to a region's instability. Recently six Persian Gulf states, for instance, formed a Gulf Cooperation Council aimed at planning "joint strategy for weapons development, exchange of experts, joint military maneuvers and cohesive armed forces development."[26]

As the NICs or different collective self-reliance systems enhance their DAP and DAP-AE potential, a follow-on step would be to turn toward the poorest LDCs in various world regions. One possibility envisioned for development and DAP that provides a possible approach for poorer LDCs is the Civil Action Military Assistance Program (CAMAP) mentioned by Benoit.[27] In developing CAMAP, he considered various relationships between different levels of spending and economic growth and found the CAMAP approach probably would produce the greatest economic benefit for a LDC. Over the time that the plan functions, defense expenditures stay at a fixed cost. Regarding weapons expenditures, 75 percent of the weapons obtained annually would come from foreign donors, while the remaining 25 percent would come from DAP. Thus, actual arms expenditures account for only 25 percent of all weapons obtained annually and most of the money spent remains in the country.

As the name civic action implies, the armed forces' involvement with the civilian community is the core of a defense spending/economic growth scheme. For Benoit, the military's interaction with the civilian community will have a tremendous spinoff capability that can immensely aid the development process. Nothing, however, has emerged in research to show if, indeed, this approach will work. Algeria is using some portions of the CAMAP approach in trying to develop both its society and its armed forces. Through conscription the government is forcing many young men to obtain more education and learn more about the outside world than most of them would do if left to their own means. Further, the Algerian army contributes to the construction of most of the country's roads, airfields, and certain lines of communication. The country does possess the capability to produce military equipment, including some small arms and ammunition, but it relies heavily upon the Soviet Union (and to a lesser degree, upon the United States) for its major military equipment needs. A full CAMAP approach, as envisioned by Benoit, may therefore contain potential for some of the poorer LDCs. This could be particularly true if these LDCs link themselves to regional NICs, rather than the more technologically developed weapons producers currently dominating the arms transfer scene.

The feasibility of any approach based upon this framework also depends somewhat on the major arms producers of the world. Ideally, some coherent and consistent multilateral agreements diminishing arms transfers to LDCs would encourage domestic production, for arms transfer cutbacks would limit the supply. However, unilateral action by one supplier (as the United States tried during the Carter administration) will not be fruitful. Other suppliers will merely take over. In turn, as is the case today, the number of willing suppliers will increase and future attempts at arms transfer control will even be more difficult than in the past. Adding to the difficulty of limiting the supply is the fact that the world's two major weapons producers, the United States and the USSR, seem oriented toward accelerating arms transfers to the Third World.[28]

Calling for weapons producers to limit AT to LDC consumers seems, therefore, to be unrealistic; an alternative to limiting the supply, of course, is to limit the demand. But to limit present LDCs' demand for military technology via AT there should be a viable replacement for the AT approach. Having a DAP program available to replace AT may be the incentive needed to really induce LDCs to cut their demands for foreign-made weapons. That the number of LDCs engaged in DAP has increased tenfold in three decades despite the expense and the lack of understanding regarding DAP's impact on a society's development could, in fact, already suggest that some countries recognize DAP is better than AT.

A Need for Convincing Research

I have argued that AT, as the current means by which most LDCs obtain their military technology, can pose problems for a developing nation's armed forces and its development process. As the tempo of the arms transfer process accelerates in a LDC so, too, will the instability in its development, thereby triggering a demand for still more arms. Applying this observation to the global system, particularly when acknowledging the rising number of militarized Third World governments, one can expect continued impairment of the LDCs' development throughout the 1980s.

In an effort to show that this would indeed be the case, one needs convincing, objective research to show how and to what degree arms transfers alter a client's armed forces and its development process. Both the South Vietnam and Iran experiences, where huge influxes of arms undoubtedly affected these nations' armed forces and their respective society's progress, provide examples for analysis. As Signe Landgren-Backstrom, a European researcher, already noted, observers perceived a problem concerning Iran's development, but no full analysis of AT effects evolved from the experience.[29]

One can say the same for South Vietnam, particularly with regard to the South Vietnamese armed forces. Massive weapons transfers, especially late in the conflict, probably altered the military's ability and approach to fighting the war and may have created more problems for both the Vietnamese armed forces and Vietnamese society than they experienced earlier. In both cases, not only did some aspects of AT most certainly work to the recipient's detriment, but the approach also boomeranged on the supplier as well. Based on these experiences, therefore, research that convincingly demonstrates the harmful effects of AT is an important first step that may move LDCs from dependence on AT to a DAP framework.

Though DAP does not initially seem a viable way to aid development, it certainly appears to be an alternative to the deleterious effects—political, social, and economic—that AT may be having on developing nations. While one cannot yet show that DAP leads to development better than production of a nonmilitary commodity, one probably can conclude that the present trend of increased AT is far less beneficial to LDC development than DAP.

Therefore, a second step in research is needed to show LDCs that a potential for better aiding their armed forces and for improving their development process exists with DAP. A DAP package, tailored within the bounds of the general framework presented earlier, should be a goal of researchers, working first with the more prosperous NICs, and then with the poorer LDCs. This acknowledgement of DAP as an important element of development also could establish criteria by which arms-producing and spinoff industries would orient themselves. Inputs from both a LDC's armed forces and the rest of its society would direct the DAP process, thereby effecting a rapprochement between the two. This constructive interaction would be an objective that certainly counters the divisiveness that AT—a foreign and often inappropriate technology—frequently fosters between the armed forces and a LDC's development process.

Urging a LDC to choose DAP over AT may be the choice of a lesser of two evils. Yet, until further analysis proves otherwise, appropriate technology as defined in this chapter needs incorporation in a comprehensive, yet flexible domestic arms-producing program for the Third World. Developing this approach to its fullest potential could see the reapplication of Ataturk's plan for developing his homeland; that is, the evolution of a LDC's armed forces ultimately would depend more upon the military technology coming from the economic, social, and political development of its society than upon outside sources little concerned with the development question.

9. Direct Satellite Broadcasting: A Case for Appropriate Technology?

Rolf T. Wigand

The communication satellite, no doubt, is the symbol of the "global village." More and more human and financial resources have been devoted over the last twenty years to ever more elaborate satellite systems. The earth is encircled by technically sophisticated satellite communication networks that are capable of reaching virtually any spot on the globe in seconds. Communication satellites have been viewed as the magic multiplier by development experts, communication scholars, and educators alike. In the early days of communication satellite development, i.e., from about 1970 to 1977, the major emphasis on using this particular technology was placed on "broadcasting" television and radio programs. Broadcasting in this context was viewed as the essential catalyst of large-scale development efforts. Since then additional emphasis has been placed on the various two-way/interactive capabilities of communication satellites, as we no longer question today's use of satellites for high-speed data transmissions or for telephone and teleconferencing services.

The use of communication satellites in the development field brought about far-reaching changes in the infrastructure of traditional communication hardware. But the impact of satellite communication on cultures and people who have yet to encounter "modernization" or "industrialization" is still unclear. Numerous questions arise in this context when communication satellites are used more and more for development in the Third World. Is high technology appropriate for reaching people to be trained in low-technology tasks? Can there be a possible justification for spending scarce resources on expensive and imported communication satellites? One also might question their place in Third World cultural and educational settings, and one could point out that they are yet another example of cultural and technological dependence resulting in increased dominance by those who own this particular technology. This chapter will discuss the advantages and potential pitfalls of communication satellites and assess their appropriateness in the development process.

Merely ten years ago countries depended on undersea cables and high frequency radio; yet today more than one hundred nations rely on communication satellites to meet their broadcast and information transmission needs. Satellites are a most desirable means to bridge long distances. Costs for terrestrial facilities are proportional to the length of the distance to be covered: the longer the distance, the higher the cost to provide the desired linkage. Satellites, however, are independent of the length of the route, i.e., connecting two facilities thirty miles apart costs as much as connecting them four thousand miles apart. Satellites therefore are cost-effective as well as efficient to bridge long distances.[1] The key advantages of communication satellites over terrestrial systems include local autonomy, flexibility, growth and coverage, improved signal quality, cost, and reliability.

In October 1957 the Soviets launched Sputnik into orbit, which marked the beginning of the space age. Since then more than 25 thousand satellites and space probes have been launched by governments as well as by various civilian organizations. Today's satellites can carry 7 thousand two-way messages or twelve simultaneous color channels. In 1960 NASA launched its ECHO I satellite into a thousand mile-high orbit. It was followed by AT&T's *Telstar* satellite in 1962 that generally was considered the first active communication satellite. *Telstar* was engineered to operate on solar power batteries. The satellites that followed required enormous ground transmitters to function effectively. Subsequently, a generation of active wideband systems could carry a television picture or its equivalent. The first geostationary satellites were launched through the Syncom program. By the time *Earlybird* was placed into orbit in 1965, artificial satellites had become a most valuable tool for the field of communication.

Transatlantic commercial transmission of television first became a reality with the Intelsat system and other satellites providing more bandwidth. In addition, it is relatively uncomplicated to connect, disconnect, and interconnect a number of disparate points when compared with conventional terrestrial systems. This versatility, no doubt, represents one of the key advantages of satellite communications since geographical barriers such as mountains or oceans can readily be overcome and, at the same time, signals can now be received by remote and less developed regions.

Communication satellites have been used successfully for voice, video, and data transmissions in numerous situations. Newspaper printing, medical care, peace missions, teleconferencing, the Olympic Games, teaching and education, as well as emergency disaster and rescue services, all have been carried out successfully via satellite communication. Today there exist at least two international systems, more than five regional systems, six military systems, and a handful of satellite systems for specific purposes such as broadcast, data relay, maritime information, and aeronautical purposes.

With the accession of Fiji, presently 102 nations are participating in the Intelsat agreement, and still more are scheduled to become signatories. At the same time, several nations have or are now planning national (domestic) communication systems via satellites. National satellite communication systems now operate in Brazil, Canada, Indonesia, Saudi Arabia, Sudan, the United States, the USSR, and, in part, Japan. Several Western European countries as well as China, India, Iran, the Middle East, and Nigeria are in advanced planning stages. Australia, Brazil, the Andean nations, and several Scandinavian countries presently are studying investment in satellites.[2] Algeria, Brazil, Norway,

Saudi Arabia, Sudan, and Zaire use rented capacity from Intelsat. The Philippine Islands use *Palapa I* rented from Indonesia. The government of India signed a contract in 1978 with Ford Aerospace for the manufacture of two satellites for its domestic system, which became operational in 1982. An Arab regional satellite system (Arabsat) is on the drawing boards.[3] An African satellite system (Afrosat) is being considered by the 38-member Panaftel group. The Andean satellite system is proposed by Aseta, a regional organization made up of Brazil, Chile, Colombia, Ecuador, and Peru. There also is the Nordic system (Nord-Sat) proposed by the five Nordic countries. For additional information, see Wigand.[4]

New generations of communication satellites presently are under construction or in the planning stages. They will establish new communication systems or expand existing ones. Numerous actors are involved in these developmental efforts. Among them are Satellite Business Systems (SBS), a partnership of Comsat General Corporation, International Business Machines, and Aetna Life and Casualty, and the Western Union—advanced Westar system in combination with the data relay and tracking system of NASA. There also is the Canadian Telesat system that launched the first commercial geosynchronous domestic communication satellite. The Federal Republic of Germany presently is building its series of TV-SAT satellites as a joint venture among the firms of AEG-Telefunken, Erno, Dornier-System, Messerschmitt-Bölkow-Blohm, and SEL. Additional organizations are referred to later on in conjunction with specific developments in the field of Direct Broadcast Satellites (DBS).

At present there is indeed no shortage of application possibilities. A satellite can *see* all of its designated surface area; distance and isolation have been removed as obstacles to the field of communication. There are very few technological hurdles that could constitute serious difficulties for future telecommunication systems.

Although there appear to be no technological difficulties, problems do arise with regard to social, legal, and policy implications. Within just a few years satellites will be broadcasting television and other signals directly into many homes around the world. One may be impressed with such technological accomplishments; many nations, however, are very much concerned about the psychological and, indirectly, sociological and political impact such broadcasts could have on their people.[5]

Since its formation in 1959, the UN Committee of the Peaceful Uses of Outer Space has been centrally concerned with these issues. This committee formed a working group to study DBS in 1968. In 1972 the USSR tabled a draft international convention before the UN General Assembly to govern direct television broadcasting. Based on this event, the General Assembly requested the Committee on Peaceful Uses of Outer Space to study the "principles governing the use by states of artificial earth satellites for direct television broadcasting, with a view to concluding an international agreement or agreements." The Working Group on Direct Broadcast Satellites and the Legal Sub-Committee of the

Outer Space Committee have been investigating these issues since 1972. Discussions and agreements have moved at a slow pace. Interested readers may want to review the *Current Report of the Legal Sub-Committee on the Work of its Eighteenth Session.*[6]

Differences with regard to philosophical, social, and cultural concerns stem largely from differences in ideologies, such as human rights and the right to freedom of information. The underlying assumption for such a position is a concrete form of freedom of expression and, consequently, free dissemination of information and ideas. Other countries, above all those of Eastern Europe, place greater importance on the right of states to protect their national sovereignty together with their social, cultural, and political values.

The delineation and relationship between free dissemination of information and ideas and the sovereign rights of states make up one of the most difficult legal issues in an effort to regulate DBS. This complicated issue readily becomes apparent when one considers freedom of expression as an absolute right. In this situation broadcasters would be entitled to have an unrestrained right to broadcast over a foreign state. On the other hand, if one subscribes to the sovereignty principle, then the consent of a state exposed to intentional broadcasting from a foreign state is mandatory. Otherwise, its sovereign rights can be infringed upon by foreign broadcasts. This area has been reviewed extensively by Dauses.[7] Various other legal aspects that may influence social, cultural, and economic conditions have been addressed by Frutkin, Galloway, and Pikus.[8]

In the following pages key developments in the field of DBS are examined. More specifically, the potential social implications of DBS are explored with respect to (*a*) the possibilities in using DBS appropriately for governmental, administrative, private, and commercial communications, and, (*b*) the potential effects and uses such possibilities may have, with particular emphasis on developing nations. These implications are discussed in terms of DBS in worldwide and regional settings, as well as for the direct-to-home and point-to-multipoint situation, word, data, and management systems, forthcoming technological developments and applications, and policy questions.

Definitions of Direct Satellite Broadcasting

Distinguishing between direct broadcasting satellites and other broadcasting satellite systems is complicated. Here, direct broadcasting by satellites is defined as those transmissions of broadcast signals via high-powered satellites that permit direct reception of television or radio signals by means of small antennas. This definition should be viewed in contrast to the type of transmission via satellites for point-to-point or point-to-multipoint systems for networking, transmission of television and radio programs, telephone service, data transfer, etc. It readily becomes apparent that a precise delineation between the two types of satellite transmissions is not quite possible. For example, a live television program could be transmitted via satellite to various regional cable television or microwave systems that then disseminate the program directly to local viewers.[9]

9. Direct Satellite Broadcasting

The International Telecommunication Union (ITU) recognized this difficulty in defining DBS. The ITU adopted the definition of the 1977 World Administrative Radio Conference, "in the broadcasting satellite service, the 'direct reception' shall encompass both individual reception and community reception." Community reception is defined as:

> the reception of emissions from a space station in the broadcast satellite service by receiving equipment, which in some cases may be complex and have antennae larger than those used for individual reception, and intended for use:
> —by a group of the general public at one location; or
> —through a distribution system covering a linked area.[10]

These definitions do not refer exclusively to physically or technically distinct concepts. Mixtures of several systems are possible as indicated above. A crucial difference between these definitions is not of a technical but of an organizational nature.[11] From this perspective it becomes important to note whether or not some controllable link exists between the satellite and the receiver.

Worldwide and Regional Communication via Direct Satellite Broadcasting

For nations to play their proper role in world affairs they must share with each other their health, agricultural, educational, cultural, and industrial knowledge. Modern transportation and communication are key factors in enhancing this activity. Until twenty years ago the regular exchange and sharing of such knowledge was difficult and cumbersome because of the vast distances involved. In addition, using various transportation methods to move people was expensive, time-consuming, and tiring. Transmitting information was slow and, depending on the circumstances, rather expensive. Today, with the development of satellites, this sharing has become fully possible, no matter how remote the location on earth may be. International meetings can be brought to local participants, possibly even into their homes, and travel no longer may be necessary. The advent of direct broadcasting via satellites has provided the opportunity for people who are separated by continents and oceans, ideological as well as language differences, and cultural discrepancies to work together to improve communication and work jointly toward common goals. The key advantage of satellite communication in this context versus its main competitors, telephone and cable, is the potential that if n earth terminals are visible from one or several satellites, then $n(n-1)/2$ two-way communication circuits exist.

Several of the experimental and demonstration satellites are prime examples of how worldwide communication was facilitated, even though—at that time— for only very specific purposes. Today 102 members belong to the International Telecommunications Consortium that launched the Intelsat series of satellites. Intelsat satellites provide communication services among major cities of the world. It is predicted that all countries will be linked via satellite by 1985. In 1978 world communication was made possible by eight Intelsat *4* and *4a* satellites providing about 80,000 voice channels via 165 earth stations in eighty countries. At present, major world capitals are linked by satellites, but a high-priority need for improved communications lies with the geographically remote and sparsely populated areas. Intelsat reports that the 1978 World Cup soccer championship was the world's biggest satellite event: a total of more than 3,200 transmission and reception hours and an estimated viewing audience of about one billion people surpassed the comparative statistics of the 1976 Olympic Games. By March 31, 1979, Intelsat functioned as the carrier for national domestic services in fifteen countries (Algeria, Brazil, Chile, Colombia, France, Malaysia, Nigeria, Norway, Oman, Peru, Saudi Arabia, Spain, Sudan, Uganda, and Zaire). Other countries that soon will join or are presently considering Intelsat's use for their domestic services are Denmark, Egypt, India, and Iraq.

Worldwide communication is also made possible by reducing the cost of needed ground stations. An example of this type of satellite was NASA's ATS-6 launched in May 1974. ATS-6 differed from previous satellites because the high transmission power was built into the satellite itself rather than into the ground station. Previously ground stations were required that cost from $100,000 to several million dollars each; now, such a ground station could be acquired for about $5,000 or less. In this way, isolated communities as well as populated areas could receive satellite signals at a much lower cost, at least when considering the specific local situation. ATS-6 was used in the United States for one year and then was given to India on loan for the next year. While the satellite was in use in India, ATS-6 successfully reached thousands of isolated villages in remote areas of that country. Programs on family planning, agriculture, health, and education were broadcast.

Eutelsat—the European consortium of seventeen member countries—has its own experimental satellite, the Orbital Test Satellite (OTS) with 6,000 voice channels. Further plans include the European Communications Satellites intended for digital communications and Eurovision in the 1980s. Other large systems are in operation or in various advanced planning stages in Canada (Telsat/Anik), the USSR (Intersputnik/Molniya-2), and Japan (BSE 14/12).

Similarly, the Federal Republic of Germany and France made their jointly financed and controlled *Symphonie 1* and *2* satellites (launched in 1974 and 1975) available to other countries interested in getting to know this technology. Egypt, India, Indonesia, Iran, and the People's Republic of China have explored the possibilities that these satellites can provide to establish modern communication services needed for economic and educational improvement as well as industrialization. Indonesia has chosen to develop its own domestic satellite system rather than rely on cable and microwave links. Cable and microwave hardware would have been too expensive as well as physically difficult to install over thousands of kilometers of oceans, hills, and forests. Two satellites, *Palapa 1* and *2*, provide new communication services to the multitude of inhabited islands in Southeast Asia.

Marisat, another communication satellite, provides ship-to-shore telephony and telex services in the Pacific. Similarly, the European Space Agency (ESA) is operating *Marots*, the Maritime Orbiting Test Satellite. An international maritime satellite organization, *Inmarsat*, has been in existence since 1979.

More than one hundred countries are now users of information from Landsat satellites. This series of satellites was designed to observe and report information on water, minerals, and ground cover as well as the earth's surface. Much of this information was previously not available. Landsat satellites provide a most useful service for national development and urban and rural planning in all parts of the world.

In 1975 the Public Service Satellite Consortium was created. Originally this organization was to serve primarily organizations in the United States, but today public service organizations from Canada and Australia have joined. The consortium was founded with the stated purpose to ensure that satellite communication and its related technology would develop and be put to use in the public service. The membership varies greatly: several prestigious universities and school districts from the states of Maine to Hawaii, the American Medical Association, the American Hospital Association, and the American College of Physicians and Surgeons. A group of other U.S. states work with professional organizations such as the American Bar Association, the American Library Association, and various religious organizations.

In Europe the interest in DBS has increased cosiderably since the 1977 WARC-ST and suggests potential applicability for developing countries. New geostationary positions and frequencies were allocated to the twenty-five European entities transmitting television programs. Several studies were undertaken to explore the possibilities provided by direct broadcasting satellites. Most European countries are consider the use of DBS (1) for economic reasons, (2) to provide for additional television programs, especially when there are limited frequencies in the VHF and UHF bands, and (3) to permit coverage of population segments with television programs in isolated and poor reception areas. Communication satellites provide easy access to reliable radio, television, and telephone services for small, isolated communities. These amenities may attract professionals, skilled workers, and technicians who are familiar with metropolitan living and conveniences to relocate into small, possibly remote communities. Access to these communication means will enhance the attractiveness of these communities and, at the same time, will enhance growth and development of industry and business within the community.

DBS could provide a desirable service to people living in such isolated and remote areas. Most city dwellers have access to broadcasting services from many stations. Many isolated and rural areas, however, are less fortunate. In the case of Canada, e.g., six million citizens live in rural areas. A large proportion of country dwellers has at best a spotty reception of AM radio. Twenty percent cannot be reached by their nearest television station. In addition, merely 35 percent are able to get FM radio. One might initially speculate that the solution to such a problem is CATV, but here the missing economic incentive makes the difference. Cable systems need about twelve to eighteen customers per kilometer to enjoy a profit. Commercial broadcasters also have a financial disincentive to enlarge their rural broadcasting service. Advertisers look for markets in the high-density urban areas. Canada has looked to other alternatives such as fiber optics in lieu of expensive copper cable. Satellite communication could improve the situation of the rural dweller by ensuring easier delivery of high-quality television reception in conjunction with community antenna systems or head ends. Government communications in remote areas are handled in many situations by private telecommunications networks and, frequently, high-frequency radio networks are being used. If the various government departments would share satellite facilities, the proliferation of private systems would cease. Such a step also would increase substantially the reliability of government communications in several remote areas of Canada. Furthermore, this switch would improve the social services offered to residents in these areas, a stated priority of the Canadian federal government. The Canadian Department of Communications is planning a direct-to-home television program delivery pilot project using *Anik B* satellite services provided by Telesat Canada.[12]

At the present time it seems unlikely that a comprehensive and integrated international DBS system will develop. Such a system could provide member nations with a full range of satellite communication services, including DBS. At one time such a system was proposed by the Twentieth Century Fund: it would have been owned internationally and would have been operated by Intelsat. Eventually, it could have been linked with Intersputnik, the Russian-proposed international satellite system, that connects Eastern European countries, Cuba, and Mongolia. It is much more likely to predict a large number of national or regional satellite systems for domestic purposes. In this organizational form, satellite systems might become—at least functionally—part of the various national broadcasting organizations. The USSR already has such a large point-to-point domestic network, and, as indicated, many other larger countries are planning satellite systems broadcasting to community receiving stations.

The present trend then is toward the development of about twenty separate systems that are global, regional, or national in nature, and not all of these will be for broadcast purposes. Each system would function as a self-contained and discrete entity. Each receiving station would be set on only those frequencies through which signals can be received from satellites in a given country's own system. This design would largely reduce problems with regard to spillover of the signal into neighboring territories.

Future plans are to provide a linkage between satellite systems without the terrestrial connection. A French earth station already has made such a linkage possible with its basic Intelsat design and the Molniya system on an experimental basis. Table 9.1 presents a listing of current and planned Third World regional and domestic satellite systems. Current and planned satellites operated and owned by international organizations or by developed countries capable of reaching remote areas are identified in table 9.2.

9. Direct Satellite Broadcasting

Direct-to-Home and Point-to-Multipoint Satellite Television Reception

Today's television signals still enter home sets from a local or regional earth-based transmitter. This way of receiving signals could change rather drastically if the cost of ground installations would be reduced by receiving signals directly from a geostationary satellite. To make low-cost ground receiving terminals feasible, the transmitter power must be increased such that the size and cost of the ground receiving installation is reduced.

The most promising, but presently hotly debated area of direct satellite broadcasting is direct-to-home television reception. This form of distribution of television programs has been discussed seriously only in Japan, North America, and Western Europe.

The Canadian Department of Communications is planning a direct-to-home television delivery project using the *Anik B* satellite under contract with Telesat Canada.[13] Canada has a long-term interest in DBS, manifesting itself in past experiments using its *Hermes* satellite. Consequently, Canada is presently not involved in direct-to-home television program delivery. In the United States a few firms such as Homesat started to offer dish antennas to private individuals. Homesat viewers can enjoy the high quality of satellite color reception and have access to a large variety of programs. The price for such a setup amounts to more than $10,000, however, an extremely high price due to the large dish antenna of 4.6 m in diameter. The price—not including shipping and installation costs—covers the following: 4.6 m dish antenna, all cable, two 12-channel amplifiers and a switch, a remote receiver that converts the satellite signal to a television picture, an adapter that adapts the television picture to the normal television set using a spare channel, frequency coordination, Federal Communications Commission license and construction permit, and pay television fees. These costs are unusually high for a large-scale direct-to-home television distribution effort. With higher-powered satellites, smaller dish antennas (69 cm in diameter) are adequate and should drastically reduce the cost of required private earth stations to a range of about $250 to $500.

Western European nations already have developed large-scale, fixed plans for direct-to-home television and radio broadcast services. Fully operational satellite systems are anticipated in the next one or two years. The European Space Agency (ESA) will serve as a carrier's carrier, and the services themselves will be marketed by Eutelsat, the consortium of Western European ministries of post, telephone, and telegraph (PTT) from Belgium, Denmark, Ireland, France, Italy, the Netherlands, Spain, Sweden, Switzerland, the United Kingdom, and the Federal Republic of Germany. This European Communications System (ECS) is intended primarily to carry television programs, but it also will be used for telephone and data transmissions.

The ESA has been engaged for a number of years in an extensive program

Table 9.1. Current and planned Third World regional and domestic satellite systems

Name of satellite system	Region/area served	Satellite system characteristics	Current status
Afrosat	Member countries of the Organization of African Unity (OAU).	Under study by the OAU, the pan-African Telecommunications Union, and ITU. AFROSAT is to be integrated with the PANAFTEL network. Still undetermined whether it will be a dedicated or leased network.	Operational in late 1980.
Arabsat (Arabian countries)	There are 22 member countries of the Arab League. At least one earth station per country is planned.	Two operational satellites at 15 E and 30 E. One spare satellite. Three-axis body stabilized; 25 channels at 6/4 GHz to deliver 8,000 simultaneous telephone circuits; one S-band CATV channel. In-orbit space craft mass 680 kg. Contractor: Aerospatiale & Ford Aerospace.	Anticipated to be operational in 1984.
Brazilsat (Brazil)	A minimum of 30 locations in Brazilian interior.	To operate in the 6/4 GHz band. The locations for two operational satellites plus spare are 60 W, 65 W, and 78 W. Contractor: still to be chosen.	Award of contract is in progress. Domestic service to 19 locations to be provided on 4 Intelsat transponder leases through 1986.
Satcol (Colombia)	Major Colombian cities and interior, as well as offshore islands.	Satellites are to operate in the 6/4 GHz band. The locations for two operational satellites plus spare are 70 W, 75 W, and 77 W. Contractor: still to be chosen.	Award of contract is in progress. Domestic service to internal network to be provided on 4 Intelsat transponder leases through 1986.
Insat 1 and 2	India and Indian offshore islands.	Two operational satellites at 75 E and 93 E. Three-axis body stabilized. Satellites will operate 12 channels of TV (12 transponders); 2 channels of those are devoted to DBS. 6/4 GHz, also 2 channels at S-band and 1 channel UHF. Contractor: Ford Aerospace.	Launched in April 1982. System will provide telephone, data, and community television network of several thousand earth terminals. Service started in 1982. Launch for INSAT 2 was planned for 1983 via the Space Shuttle.
Iscom (also referred to as Apple India)	Experimental project for India (French Ariane Passenger Pay Load Experiment).	One satellite launched on test Ariane 3 with limited communications capacity. Designed by India Space Research Organization.	Launched in 1981. Solar array deployment problem limited effective capacity of this satellite.

System	Coverage	Characteristics	Status
Palapa A and B (Indonesia)	Indonesia and some service to ASEAN countries of Malaysia, Philippines, Singapore, and Thailand.	Two operational PALAPA A satellites are situated at 77 W, 83 E. 12 channels in 6/4 GHz. Weight in orbit is 294 kg. H-333 design built by Hughes. Two PALAPA B satellites plus spare are located at 108 E, 113 E, and 118 E. 24 channels in 6/4 GHz, HS-376 design with 8-year lifetime expectancy. Weight in orbit is 638 kg. Contractor: Hughes.	Two PALAPA A satellites were launched in 1976–77 and are operational with a network of 40 earth stations. Two PALAPA B satellites were launched in 1982. A large number of 4–5 m earth stations (18–22 db/k) were deployed for PALAPA B.
Satmex (Mexico)	Telephone, data, and television distribution to remote areas of Mexico.	One SATMEX satellite to operate in 6/4 GHz band. Location will be 85 W. Contractor: still to be chosen.	SATMEX is to be launched in the mid- to late 1980s. RFP not yet issued. Domestic service to a network of more than 100 earth stations will be provided through the lease of 6 Intelsat transponders through 1986.
P.R.C. (People's Republic of China)	China's major cities and remote areas.	The PRC-1, 2, and 3 satellites are to operate at the locations of 65 E, 80 E, and 95 E. They are to operate in both the 6/4 GHz and 14/11 GHz bands.	The PRC system has been postponed indefinitely. A lease of one or two Intelsat transponders is to provide interim service until a decision is reached.
STW 1 and 2 (People's Republic of China)	China's major cities and interior.	These two satellites are to be operational at 70 E and 125 E. STW 1 and 2 are low-capacity experimental communication satellites.	Launch dates cannot be determined at this time.

Table 9.2. Satellites operated and owned by international organizations or by developed countries capable of reaching remote areas

Name of satellite system	Using agency/ owner country	Remote area served/purpose (if known)	Current status
Anik	Telsat (Canada)	Nothern Canada	in operation
ATS-1	NASA/AID; University of South Pacific and Peacesat (United States of America)	Micronesia and Southwest Pacific	in operation (to be terminated since its useful lifetime has been exceeded)
Aussat	OTC (Australia)	Australian remote areas and Papua New Guinea	planned
CBSS	Department of Communications, Government of Canada (Canada)	Northern Canada (DBS service)	planned
Comstar	AT&T/GTE (USA)	Alaska	in operation
Ekran	Government (USSR)	remote areas of USSR	in operation
European Communications Satellite (ECS)	Eutelsat (Western European countries)	North Africa/Middle East (television service only)	planned
Gorizont	Government (USSR)	USSR remote regions (television service mainly)	in operation
Intelsat	Intelsat, Inmarsat (106 countries)	Atlantic, Pacific, and Indian oceans and all land areas	in operation
Loutch	Government (USSR)	USSR remote areas	planned
Marisat	Comsat General (USA)	Offshore drilling rigs	in operation
Marecs	Eutelsat-leased to Inmarsat (European countries)	Offshore drilling rigs in ocean areas	planned
Molniya	Intersputnik (USSR)	Asia, Africa	in operation
Raduga	Intersputnik (USSR)	Asia, Africa	planned
RCA Satcom	RCA/Alascom (USA)	Alaska	in operation
Stationar	Intersputnik (USSR)	Asia, Africa	in operation
STC (Satellite Television Corporation)	Satellite Television Corporation (USA)	Rural United States (DBS service exclusively)	planned
Symphonie	French and German governments (France and Federal Republic of Germany)	Africa, Atlantic ocean	in operation (experimental/ testing period)
TDRSS (Tracking & Data Relay Satellite	NASA (USA)	Ocean regions of the world (data relay)	planned

Table 9.2. (continued)

Name of satellite system	Using agency/ owner country	Remote area served/purpose (if known)	Current status
Telecom	D.G.T. (France)	French overseas territory	planned
Tele-X	Government (Sweden)	Remote regions of Sweden and other Nordic countries	in operation
Volna	Government (USSR)	Offshore drilling rigs (Pacific and Indian oceans)	in operation
Westar	Western Union (USA)	Alaska	in operation

of research and development that has strengthened considerably European-based space technology. European firms such as West Germany's Messerschmitt-Bölkow-Blohm (MBB) have been negotiating with several European and Asiatic countries for the production of satellites. During the spring of 1979 China signed an agreement with MBB for a joint production program to deliver a total of seven television satellites. In the past France and Germany have cooperated in the development of *Symphonie*, an experimental satellite, that is in operation in several programs over the Atlantic and Indian oceans. *Symphonie* is used to link France with French-speaking areas in Africa, the Carribean, and North America. Another joint venture by France and West Germany was scheduled for 1983 to provide direct television in the home.

France's proposed *Telecom-1* will serve metropolitan France (including Corsica) and her overseas *Départements* (Guadeloupe, Guyana, Martinique, Mayotte, Réunion, and Saint Pierre et Miquelon). *Telecom-1*—with a second satellite on a standby basis—will provide television and telephone transmissions, but also digital wideband, high-speed connections between 3 m dishes installed at the premises of major users. *Telecom-1* will have two main objectives: (1) A service providing "links with overseas *Départements*" for handling traffic (telephone and television). (2) An "intra-company link" service providing digital wideband, high-speed links between different branches of an organization. Initially *Telecom-1* will serve well-defined users, i.e., the cost of earth stations will be within the reach of large-scale users only. This satellite—via the use of mobile earth stations—will provide temporary links for the transmission of television programs. *Telecom-1* will be launched by means of the European launcher Ariane and will be the first of a new system of European satellites to be placed into orbit during the second half of the 1980s.[14]

Direct broadcasting satellites reach entire nations and connect peoples and continents. In the case of Europe, DBS does not lend itself well for local or regional programs, neither for feedback programs nor those that require audience participation. This is due, in part, to the close proximity of neighboring countries with their own channel allocations. In the case of North America, for example, DBS lends itself ideally to overcome large distances and—by proper delay—time zones. Local and regional programming and transmission become desirable, feedback and audience participation become possible and have been used successfully in past demonstration experiments.

The Future of Satellite Communication

New DBS-based communication networks are planned that promise to greatly extend the speed, reach, and reaction time of users. The following plans of several organizations will give some indication of what to expect in the future.

There is no doubt that technology—and DBS in particular—will reduce the importance of location. It may make it easier to start a business enterprise and may encourage competition. There are definite indications that top management during the next twenty-five years may be *remote management*, and these managers will not have a need to see their subordinates on a daily basis. Subordinates can be reached by the touch of a button. The advent of satellites overcomes geography, and many persons will be able to work in areas physically remote from their office or factory. Generally, such newer management practices via satellites will facilitate the size of an organization. Widespread information systems will provide instant information retrieval and communication. Flexibility no longer will be an issue as it is known today. Specialization is bound to increase in conjunction with the above-described technological developments. This specialization may require more work and career changes, and it may place additional emphasis on continuing education, training, and development in these communications technologies.

The rapidly increasing and expanding use of satellites in conjunction with cable, microwave, computers, and other technologies is drastically changing the uses and applications of broadcasting. In many cases the advent of satellite-related technology can free the user of the constraints of previous life-styles. Homes, schools, and hospitals, as well as businesses and governments, are gaining access to numerous channels for a multitude of services. Satellites can widen opportunities for television program distribution by linking up with diverse media to meet joint functional and geographical needs. It appears that satellites may have become the universal link for broadcasting and related technologies in a wireless world.

New applications for direct broadcast satellites have been explored by NASA through its series of Applications Technology Satellites (ATS) 1, 3, 5, and 6, the joint NASA and Canadian Department of Communications CTS experimental and commercial satellites.[15] The main new and promising DBS applications are direct broadcast of television and radio, tele-education, telemedicine, transportable emergency terminals, simultaneous multi-language translation between two continents, high-speed facsimile transmissions, electronic mail, electronic newspapers (Videotext), and electronic shopping and funds transfer. Satellite communication is a prime example of technology advancing more rapidly than concurrent uses and applications in terms of political constraints, ideological issues, legal complications, and economic considerations. The applications referred to earlier as well as others are not dependent on future breakthroughs in technology. Today's satellite technology is fully operational and practical, using existing terminals and sender and receiver stations.

Many other application possibilities exist. In 1974 the Canadian Broadcasting Corporation used Telesat to demonstrate how modifications in remote television stations would allow radio programs originating in the south—or any other location—to be received simultaneously with television signals. Telesat's transportable earth stations allow for a maximum of flexibility. They allow broadcasters to disseminate live television programs from locations without permanent transmission facilities.[16]

A good number of other experiments already have demonstrated the potential of providing various types of message services such as teletype, telephony, and facsimile and data transmission in numerous settings ranging from emergency and disaster situations to offshore oil rigs. One of the major uses of the new Anik-C series of Telesat, Canada's domestic satellites, is designated for digital voice communications traffic between ten major Canadian cities.[17]

Technological Developments in Space

Research scientists soon will be able to carry out much of their work directly in space. Numerous scientists already are predicting that with the advent of the Space Shuttle as well as new technology in the 1980s will make possible the construction, assembly, placement, checkout, repair, maintenance, retrieval, and stabilization of satellites in space.[18] Others are planning on the construction of large space platforms for building structures fifty to seventy meters long, so-called orbital antenna farms.[19] The key to all of these endeavors is the Space Shuttle, NASA's space transportation system. The present trend is somewhat of an inversion of past practices in that previously scientists made every effort to produce satellites as small and as cost-efficient as possible; current trends are increasing satellite antenna size and power. Using the old type of satellite implied that very large and expensive earth stations had to be built. The new generation of satellites with increased antenna size, power, and sophistication, however, makes it possible to reduce the power and antenna size of the earth terminals, which leads to lower construction and operating costs. These savings —at the same time—will make it possible to build more earth stations for whatever purposes. The future—at least from the distribution, consumption, and technical side—of DBS indeed looks bright: experiments have beamed color television through earth dish antennas as small as 0.6 m. Presently a receiver station as small as 1 m costs under $1,000.[20] The West German-produced receiver stations with a diameter of 0.69 m are said to cost between $280 and $560 when mass-produced.

There are a number of other technological developments that will alter the use of satellites. Among these innovations are the use of multibeam signals as well as on-board switching capabilities that can be expected with the third generation of satellites. These third generation satellites will be planned during the 1983 to 1986 period. Considering parallel developments at the present time, it does not seem very likely that all the various kinds of social, political, legal, technical, economic, and other problem areas will be solved satisfactorily so that satellites can be used at an international level to their full capabilities. Taylor[21] and others speculate that it will be the fourth generation of satellites in the mid-1990s that eventually will put satellites to their full use.

This fourth generation of satellites has aroused considerable interest. It is largely believed that this generation will be of a very different kind of satellite than are presently known. Such structures as orbital antenna farms are too large to be launched in their final shape. Various parts and pieces will be transported by several flights via the Space Shuttle and will be assembled in space. Maintenance will occur either by an on-board crew or by regular visits of maintenance personnel. The Space Shuttle will have reached parking orbit 300 km, or 182 miles, above earth, and Edelson and Morgan[22] assert that the cargo bay doors would open and the various prefabricated sections of the space platform then could be connected. A cranelike structure, a "cherry picker," will aid in this process.[23] This structure would allow for the mounting of antennas with a diameter of up to 33 m together with smaller antennas, transponders, switchers, and other equipment. Most of the equipment would be modular to facilitate maintenance and repair as well as making easier the addition of equipment to the platform. Such large platforms would have numerous advantages: economies of scale and maintenance, for example, can be shared by all services on the platform; there can be interconnections between systems; launch costs per transponder can be reduced; interference caused by crowding in orbit can be lessened; smaller earth stations are possible; savings can be realized due to sharing of antennas, power, and equipment; and there can be an increase in flexibility, among other things.

Edelson and Morgan[24] describe one such gigantic platform that could fill all the communication needs for the Western Hemisphere. The capacity of such a proposed antenna farm is most impressive: it would include seventeen separate systems, would generate 20 kW, and would be the equivalent (in bandwidth) of 927 transponders as they are known today. Only eighteen of these are designated for television distribution, only one for direct-to-home broadcasting, and only one-half of one for educational television. Expressed differently, such an antenna farm would be the equivalent of seventy-seven Westar-type satellites. Such a setup could be the basis for an all-satellite broadcasting system that would fulfill every imaginable need.

Terrestrial DBS Applications

Since about 1976 a number of researchers have developed rather attractive satellite system concepts. Recently, Bekey[25] and his coworkers under contract with NASA have developed a number of conceptualizations ranging from personal radio-telephone terminals the size of a wristwatch to holographic image transmission for teleconferencing via satellite and small rooftop antennas. Bekey's ideas are the results of the implications of the complexity inversion technique, i.e., the shift from the past practice of small satellites and large, expensive earth terminals toward the future direction of satellite technology involving larger and more powerful satellites with very small and inexpensive earth antennas. The implications of these developments are impressive: rather than necessitating "wired cities," this complexity inversion would create "wireless cities." DBS would then become a potentially very effective and affordable means for developing nations to participate in the forefront of satellite communication.[26] To explicate these developments, three examples will suffice here: personal radio-telephone terminals, electronic transmission of mail, and the wide dissemination of educational television.

Personal radio-telephone terminals are part of the personal communication satellite system proposed by Bekey.[27] Its aims are to interconnect about 10 percent of the 1990 United States population, which would be about 25 million people. This interconnection would become possible through wrist-watch-sized radio-telephones, and communication would occur on a direct user-to-user basis. Users could be in one particular location or mobile, and communication with anyone else would become possible via a ground-terminal entry point. Without going into highly technical details, it will suffice to note that rechargeable batteries with enough energy to transmit at least five one-minute calls per day without recharging are considered adequate. The personal radio-telephone terminal features a microphone and loudspeaker, an antenna, a pushbutton dialer, an emergency override button, a transmit button, and a number display. This device would weigh no more than a large wristwatch. Bekey estimates that such a terminal could be mass-produced for about $10. A satellite-based discipline would impose a call duration of one minute, and requests for longer calls would receive a lower priority in the queue of calls received. Emergency calls would immediately receive highest priority. Bekey states that developments toward such a system could readily permit commercial operation in 1990. The entire space-based system is estimated to cost merely $2.3 billion as compared to its terrestrial equivalent of $20.7 billion. In comparison and for the same number of channels, today's telephone system charges about $13.8 billion. Such a satellite-based system allowing direct reception would enjoy a wider coverage than the present microwave/wire-based terrestrial systems. Bekey estimates that the total research and development costs, investments, and operations costs could be amortized within a ten-year period by charging its customers merely $0.036 per minute. Costs for any terrestrial system would increase with increased long distance calls. Long distance, once again, is no longer an issue for satellite transmissions of signals.

Bekey's electronic mail system would exchange first class mail between business and government buildings via satellite. Direct home reception is not discussed in his conceptualizations. Such a system could deliver about 15 million pieces of mail per year among 544 thousand small terminals, each installed on top of a separate office complex, with corresponding hard-copy receiving devices. Fifteen million pieces of mail reflects about 15 percent of the anticipated 1990 U.S. mail flow. Initial users would be corporations with assets of more than $1 million. All messages would be digital, and each would include a receiver-sender code number. Each terminal is designed to have a ten-minute buffer storage. The United States Postal Service as well as its counterparts in several European countries have engaged in preliminary studies and experiments involving the transmission of electronic mail. These systems, however, are still dependent on the physical pickup and delivery of the mail piece. Bekey's proposed system requires only an in-house distribution. A cost comparison revealed that the USPS hybrid system could deliver a letter for $0.099 per letter, and the satellite-based system proposed by Bekey could do the same for merely $0.026 per letter. The mere cost comparison may suggest the possibility of rushing ahead with a pure space-operated system. On the other hand, the various social implications are still difficult to foresee in their entirety. Policy makers will have to struggle with job displacement of a large working force and retraining this working force for other jobs. Still other problem areas are that the same organizational structure will have to remain for those individuals who cannot be reached by electronic mail. Such exclusions will occur by design or because of package delivery, other classes of mail, etc.

The educational television system proposed by Bekey would interconnect all 65,000 schools in the United States as well as their 16,000 district headquarters (or all 4,000 universities and colleges with 250,000 remote learning sites) with color television and interactive audio. Such a system via the use of satellites would use small dish antennas on each school building to participate and disseminate its own messages and programs. At the same time programs could be shared between poor as well as rich districts throughout the United States. To make such an effort feasible, the satellite would have to transmit almost 1,500 television channels simultaneously through 600 beams, each covering an area of 80 by 160 km on earth. A total of 634 simultaneous "uplink" channels would be required from the districts for program origination. The satellite would transmit its signals to 0.92 m antennas installed at each school. Transmission from the schools would occur via antennas with a diameter of 3.05 m located on school district buildings. Bekey made a cost comparison on the basis of a six-hour day for instructional programming five days a week, for nine months per year, with thirty students per class, and half the classrooms (about ten rooms per school)

Table 9.3. User cost comparison for educational television

System	Cost per school per year[a]	Cost per classroom hour in dollars	Cost per pupil	Total 20-year costs in billions of dollars
Space initiative	$ 4,200	$ 0.36	1.2¢	$ 2.73
Ground alternative	20,800	1.78	5.9	13.53
Teachers	116,000	10.00	33.3	75.4

a. Average of 10 classrooms per school equipped for educational television; 6 hours a day, 5 days a week, 9 months each year; 30 pupils per classroom.

equipped for educational television in each of the 65,000 schools in the United States. The space initiative proves to be far the least expensive as reflected in table 9.3.

These figures are compared against the average cost for a teacher, which resulted in about $10 per hour. Assuming that the ten classrooms would have simultaneous instruction, the resulting cost of teachers per school per year would amount to at least $116,000 as compared to $4,200 for the space initiative and $20,000 for the strictly terrestrial alternative.

Policy Implications

Given virtually unlimited technical possibilities in DBS, it appears that the information and communication industry must participate actively in these technological developments. This point becomes important considering that broadcasters and user groups, for example, are told what they can do with the satellites after the fact; they deserve to have an active part in these future developments. In a sense it becomes bothersome to note that the WARC 1979 conference in Geneva attempted to allocate the airwaves up to the year 2000. How can proper allocations be made that are functional when the various technological developments of satellite technology are not yet determined? Taylor[28] points out that this implies putting technological development in a strait-jacket for twenty years when we have no idea what services the next generation of viewers will want.

Satellites are at present the essential element to broadcast and other information dissemination development until the year 2000. Some of the technological developments described above may appear today as science fiction dreams. These developments, however, are good possibilities for the future. The technology in many instances already exists, but its implementation in many cases means a large-scale effort, and there may be doubt if some of these ideas are economically justified at present. There are other hurdles to be overcome, such as international agreements, national regulatory boards and agencies, as well as various governmental bodies. The advent of satellite broadcasting is a primary example of how technology has developed well before humankind is able to put it to proper use when considering social, legal, and international implications.

A good number of important questions will have to be answered that are, in part, based on the organizational control of the airwaves in various countries (such as the issuing of licenses to private operators by the government) versus the functioning of a monopolylike organization in the form of the PTTs. For example, will the PTTs be competing with private communication companies? Will the European PTTs invest their funds in cable networks because of previous heavy investment efforts in terrestrial systems, even though a rooftop-to-rooftop, satellite-based system may be cheaper? Is the lack of competition in the satellite business the best route to go for PTTs, and is this in the best interest of the public? DBS may trigger a long overdue reevaluation of the various social needs and gratifications as well as national purposes that television, radio, and other information outlets serve. The needs differ from country to country, largely based on their levels of development in several areas. The final test, however, will be whether numerous and diverse institutions from the private and governmental sectors for broadcasting, telecommunications, and satellite services eventually can cooperate and act in concert. Only when these institutions have adapted this new technology can national information and broadcasting goals be achieved.

One of the main advantages of DBS is its possibility of reaching a worldwide audience, perhaps thereby enhancing worldwide understanding and cooperation. This is not to say that this audience would be watching the same global television program at the same time. There are too many factors that would not make such an effort feasible: language barriers and cultural differences would fragment a worldwide audience; one-third of the world is asleep at any given time. There are only a few truly international events where such a network would benefit a worldwide audience: the Olympics, soccer championships, world crisis situations, space explorations, UN events, for example. This limited demand, however, can be met by interconnecting various national and regional broadcast satellite systems.

In developed and industrialized countries, DBS is likely to be nested into existing networks in an auxiliary role. This seems to be happening and will be the emphasis for the very near future. In the case of the United States, the main interest in satellites has been to provide cheaper networking facilities in lieu of coaxial or microwave terrestrial connections. Especially in the case of Western Europe, today's frequency assignments do not make possible more than four national television programs in each country. As of now no Western European country has developed more than three television channels (programs). Past television viewer's behavior in border areas, however, has demonstrated that viewers are willing to pay for sets and antennas that can receive programs from neighbor countries.[29] This behavior may be an indication that the demand for additional programs has not been saturated. DBS could meet such a demand for Western Europe.

Efforts at the national and international levels are necessary to see what roles DBS can play in a given country's development process. In the past, developing countries designed their information and broadcast systems such that they reach the people in major urban areas; they reach the rich, the educated, whereas the accessibility to the media and other information outlets in many areas is limited.[30] Content and form of the messages usually are tailored for urbanites. Messages become dysfunctional when content and form do not agree with the cognitive structure of the recipient.

Developing countries are fully aware of the important role telecommunications —of which DBS may be viewed as a subset—can play in the development process. The 1979 World Administrative Radio Conference was just one international setting in which the less developed countries stressed these concerns in the form of a series of resolutions. These resolutions included the following:

—technical cooperation in national propagation studies in tropical areas designed to improve and develop the developing countries' radio communications;
—transfer of technology in telecommunications for the purpose of developing services and attaining social, economic, and cultural objectives of the developing countries;
—promotion of telecommunications in rural development for education, health, agriculture, and other activities important for social and economic progress;
—international cooperation and technical assistance in space radio communication, with the goal of making available any means of technical assistance in space communications to less developed countries.[31]

DBS lends itself better than most conventional communication means for serving this important role in the development process.

The benefits of DBS for developing countries are manifold. Among them are: improvement of secondary (manufacturing) and the tertiary (government, finance, and services) sectors; potential substitution for travel; potential energy savings; decentralization of business and industry through the capability to transfer information in a timely and accurate fashion; providing information to consumers and facilitating accurate ordering and delivery of goods; maintenance and expansion of tourism, which in turn expands the service sector; increased efficiency, cost effectiveness, equitable distribution, and geographic coverage for government administration and delivery of services; organizational impacts on agricultural production through improvements in ordering and delivery of supplies and equipment; and more timely access to services and increased availability of marketing information.

The advent and possibility of widespread use of DBS gives policy makers another chance to examine the *structure and function* of their respective information and broadcast systems.Such dysfunctions have been experienced with instructional television, as well as with health care delivery systems using DBS.

In part, the problem lies in the design of the telecommunication service that quite frequently does not fit the pattern of instruction or health care. This implies that the users must take the initiative to use DBS in their setting and shape it to their needs. Similarly, policy makers in most developing countries previously started and implemented information and broadcast policies with an overwhelming emphasis on structure. This means that broadcast systems typically were initiated in highly populated urban areas. These residents received radio and television content that many times was ill-suited for their local and national purposes. In addition, even when the content would have been well-suited for the urban dweller, it may have been the case that the rural resident would have been in higher need of information and broadcasting services in terms of national development priorities. It can be seen readily that a proper interplay between structure and function when designing a broadcast system did not occur in many developing countries, and a *structural* and then a functional approach was typically emphasized. In such a setting, mass media are socializing agents in the development process. With the advent and possible widespread use of low-cost DBS even for developing countries, policy makers should emphasize a functional-structural approach, where one starts with the functions of information and a broadcast system from which structure will emerge. DBS readily overcomes the previous structural hurdles with regard to transmitters, geographical problems, etc., since DBS can be available anywhere in a wireless world.

Truly integrated features of satellite-based communication networks demonstrate the importance of providing integrated networks for development planning purposes in less developed countries. Such integrated communications planning rarely occurs. Typically, communication investment policies in less developed countries recommend that social and economic activities are to be supported only after these activities are well established. Remote and rural telecommunication systems must be planned as integrated systems and must become an integral part of rural development efforts. Only through such rigorous and integrated communications planning is it possible to have the desired social participation and make added risk-taking behavior possible. Such an integrated, satellite-based network can become a powerful tool for speeding up considerably economic and social change multiplier effects, as well as for the transfer of technology on a national basis. Various social and economic demands can be combined in this fashion and make national planning possible in the first place. Villagers can then interact, jointly participate with urban dwellers, and thus achieve national economic and social goals.

Conclusions

In developed and industrialized countries satellites allow broadcasters to operate as they have in the past. What is new is the added convenience, the lower costs, the realization that distance is no longer a concern, the increase in the number of broadcast sources, as well as flexibility and immediacy. There is a widely held belief that broadcasting via satellites will establish a new and more vital image for the television medium. Especially in the case of less developed countries, the possibility exists of reaching large audiences at a low cost and with tailored programs and information essential to their future development and survival in a complex and ever-changing world.

Presently many developing countries do not operate extensive television information and broadcast systems. The advent of DBS, no doubt, constitutes a potential of a magnitude that is difficult to assess today. At a relatively low cost, entire national information and broadcast systems could be established without the high costs of terrestrial interconnections. Economic realities would at first not allow for direct-to-home reception, but community receivers are likely to be installed by the states' governments.

In the case of developing countries, one of the most difficult problems to overcome in satellite transmissions is not at all related to overcoming long distance communication. This particular problem is referred to as the "last mile problem," referring to the means by which the message is finally transmitted to the actual user. One way of overcoming this difficulty is via the use of community or village receiving devices (e.g., large-screen television sets). The actual in-home delivery of such information on an individual basis constitutes the central issue of this "last mile problem." One could argue that the "last mile problem" may be the key missing element in DBS necessary for social and economic development in remote and rural areas.

It is obvious that in this situation access to the system and control over programming belong to the country's government and are subject to its priorities. In most developing countries education of citizens ranks as the top priority. Wigand[32] and others have described the value of television programming for national development by providing education and information as well as a sense of national identity and unity. Past demonstration experiments already have shown that satellite broadcasting holds substantial promise of accelerating national development for educational purposes in such areas as literacy, family planning, health care, training in agricultural and mechanical skills, and others.

Decisions for the implementation of such systems, however, will have to compete with other priorities. Often the engineering and technical know-how will have to be imported for the design and operation of such systems. The old problem in regard to indigenous programming most likely will remain, and much of the software may have to be acquired abroad just as in the past. If the latter prevails, translation and cultural differences of educational programs may reduce the communication effectiveness, if not constitute a barrier.

These brief deliberations suggest that DBS may not necessarily provide the panacea to fulfill the educational and cultural needs of developing countries in the next few years. Bilateral, if not multilateral, cooperation as well as regional or interregional efforts will be necessary to accomplish these goals.

Notes

1. What Technology Is Appropriate?

1. Richard S. Eckaus, *Appropriate Technologies for Developing Countries* (Washington, D.C.: National Academy of Sciences, 1977), pp. 37-52.
2. Nicolas Jaquier, ed., *Appropriate Technology, Problems and Promises* (Paris: Organization for Economic Cooperation and Development, 1976), p. 175.
3. E. F. Schumacher, *Small is Beautiful—Economics as if People Mattered* (London: Blond and Briggs, 1973).
4. Eckaus, *Appropriate Technologies*, pp. 48-50.
5. United Nations Development Program, *Employment Incomes and Equality—A Strategy for Increasing Productive Employment in Kenya* (Geneva: International Labor Office, 1972), p. 135.
6. C. P. Timmer, J. W. Thomas, L. T. Welk, and D. Morawetz, *The Choice of Technology in Developing Countries* (Cambridge, Mass.: Center for International Affairs, Harvard University, 1975), pp. 3-15.
7. Stephen S. Rosefeld, "Sudan—A Development Revolution," *Washington Post*, December 1, 1978.
8. R. J. Congden, ed., *Introduction to Appropriate Technology* (Emmaus, Pa.: Rodale Press, 1977), pp. 117-32.
9. Jaquier, *Appropriate Technology*, pp. 260-75, 286-95.
10. Congden, *Introduction to Appropriate Technology*, chap. 3.
11. Francis Stewart, *Technology and Underdevelopment* (Boulder, Colo.: Westview Press, 1977), chap. 10.
12. Ibid., chap. 9.
13. Congden, *Introduction to Appropriate Technology*, pp. 80-101.
14. Jaquier, *Appropriate Technology*, pp. 296-308.
15. Timmer et al., *The Choice of Technology*, p. 73.
16. "The Choice of Technology in Developing Countries," *World Development* 5, no. 9/10 (September/October 1977): 773-882.
17. United Nations Development Program, *Employment Incomes and Equality*, pp. 371-82.
18. "Transportation Technology Support Project," *Transportation Research News*, no. 78 (September/October 1978): 10-11.
19. M. Allal, G. A. Edmonds, and A. S. Bhalla, *Manual on the Planning of Labor Intensive Road Construction* (Geneva: International Labor Office, 1977).
20. Jack Baranson, "Road Construction Equipment for Local Manufacturer—Kenya," *Ekistics* (June 1977): 369-72.

2. Intermediate Technology and Development

1. E. F. Schumacher, *Small Is Beautiful: A Study of Economics as if People Mattered* (London: Blond and Briggs, 1973).
2. The background and development of Schumacher's thoughts have been captured by his long-time colleague and friend, George McRobie, in his book, *Small Is Possible* (New York: Harper & Row, 1981).
3. E. F. Schumacher, "The Problem of Unemployment in India," an address at the India Development Group (London, June 1971).
4. See, for example, the following annotated bibliographies: Marilyn Carr, *Economically Appropriate Technologies for Developing Countries* (London: IT Publications, 1976); Gareth Jenkins, *Non-Agricultural Choice of Technique*, (Oxford: Institute of Commonwealth Studies, 1975); and David French, *Appropriate Technology in Social Context* (Washington, D.C.: USAID, 1977).
5. G. A. Edmonds, and J. D. F. G. Howe, eds., *Roads and Resources: Appropriate Technology in Road Construction and Developing Countries* (London: IT Publications, 1980); also William Armstrong, *Better Tools for the Job: Specifications for Hand Tools and Equipment* (London: IT Publications, 1980).

6. Derek Miles, *A Manual on Building Maintenance*, vol. 1 (Management) and vol. 2 (Methods) (London: IT Publications, 1976); also Derek Miles, *The Small Building Contractor and the Client* (London: IT Publications, 1980).

7. *Report on the Activities of the Building Materials Workshop, Cradley Heath* (London: IT Publications, 1978).

8. Jon Sigurdson, *Small Scale Cement Plants* (London: IT Publications, 1977).

9. Peter Fraenkel, *Food From Windmills* (London: IT Publications, 1975).

10. McRobie, *Small Is Possible*.

11. John Davis, *Technology for a Changing World*, papers edited by Roger England (London: IT Publications, 1978).

Suggested Further Reading

J. Davis, *Technology for a Changing World*, papers edited by Roger England (London: Intermediate Technology Publications Ltd., 1978).

P. D. Dunn, *Appropriate Technology: Technology with a Human Face* (London: Macmillan, 1978).

D. French, *Appropriate Technology in a Social Context* (Washington, D.C.: VITA, 1977).

D. H. Frost, "Appropriate Industrial Technology: An Integrated Approach," in United Nations Industrial Development Organization, *Conceptual and Policy Framework for Appropriate Industrial Technology*, Monographs on Appropriate Technology No. 1 (New York: United Nations, 1979).

M. Harper and Tan Thian Soon, *Small Enterprises in Developing Countries* (London: IT Publications, 1979).

N. Jequier, ed., *Appropriate Technology: Problems and Promises* (Paris: Development Centre of OECD, 1976).

G. McRobie, *Small Is Possible* (New York: Harper & Row, 1981).

E. F. Schumacher, *Small Is Beautiful* (London: Blond and Briggs, 1973).

F. Stewart, *Technology and Underdevelopment* (2nd Ed.; London: Macmillan, 1978).

P. Timmer et al., *The Choice of Technology in Developing Countries: Some Cautionary Tales* (Cambridge, Mass.: Harvard University Press, 1975).

3. The Political Economy of Intermediate and Appropriate Technology

Thanks are due to my colleagues Richard K. Ashley, Mathew J. Betz, and Rolf T. Wigand for comments and criticisms of earlier versions of this chapter.

1. This quasi-religious vision is most clear in E. F. Schumacher, *A Guide for the Perplexed* (New York: Harper & Row, 1977), but it also surfaces in his better known *Small Is Beautiful: Economics as if People Mattered* (New York: Harper & Row, 1973), especially "Buddhist Economics," pp. 53–62.

2. By "intermediate technology" I mean tools, machines, workplaces, and ideas (blueprints) where the capital investment per worker is in the range of hundreds of dollars rather than many thousands of dollars found in advanced technologies and the few dollars per worker of traditional techniques. Appropriate technology covers the entire possible spectrum of investment per worker. For Schumacher and his followers, appropriate technology is based upon the principles of smallness, simplicity, capital saving, and nonviolence toward people and the environment. This is probably the correct mix for rural development in LDCs. However, in this book we take a wider view of appropriate technology that emphasizes multiple criteria of technological choice in addition to economic efficiency (see chapter 1 by Mathew Betz).

3. Besides the chapters in part 2 of this book, applications of intermediate and appropriate technology are extensively discussed in Donald D. Evans and Laurie Nogg Adler, eds., *Appropriate Technology for Development: A Discussion and Case Histories* (Boulder: Westview Press, 1979), and George McRobie, *Small Is Possible* (New York: Harper & Row, 1981).

4. See Werner J. Feld's chapter, "The Transfer of Technology to Third World Countries: Political Problems and International Ramifications," in this book.

5. See Immanuel Wallerstein, *The Modern World-System: Capitalist Agriculture and the Origins of the Capitalist World-Economy in the Sixteenth Century* (New York: Academic Press, 1974); Wallerstein, *The Modern World-System II: Mercantilism and the Consolidation of the European World-Economy, 1600–1750* (New York: Academic Press, 1980); and Wallerstein, "The Rise and

Future Demise of the World Capitalist System: Concepts for Comparative Analysis," *Comparative Studies in Society and History* 16, (September 1974): 387–415. Braudel's masterwork is F. Braudel, *The Mediterranean and the Mediterranean World in the Age of Philip II*, 2 vols. (New York: Harper & Row, 1976).

6. The following discussion draws from Christopher Chase-Dunn and Richard Rubinson, "Toward a Structural Perspective on the World System," *Politics and Society* 7 (1977): 453–476. The "world" included within the MWS at any time is not equal to the entire globe. Only in the late nineteenth century did the two become equivalent.

7. Whether or not this revival of international trade was a cause or an effect of the emergence of capitalism has been subject to a lively debate. See Rodney Hilton, ed., *The Transition from Feudalism to Capitalism* (London: NLB Verso, 1978).

8. Some LDC workers do receive high wages and employ advanced technologies. Most often this is true in mining and petroleum, but it sometimes occurs in manufacturing. More often than not, however, these activities are controlled by transnational corporations and are thus an extension of a core mode of production into the periphery. In a few instances high wages and advanced technology in an LDC are found in a state enterprise. This is so because, with but a few exceptions in LDCs, only the state can amass sufficient capital to employ the latest technologies.

9. Frederic C. Lane, *Profits from Power: Readings in Protection, Rent and Violence-Controlling Enterprises* (Albany: State University Press of New York, 1979).

10. Quote in Andrew Robertson, "Introduction: Technological Innovations and Their Social Impacts," *International Social Science Journal* 33 (1981): 436.

11. On state building in comparative historical perspective see Charles Tilley, ed., *The Formation of National States in Western Europe* (Princeton: Princeton University Press, 1975), and Perry Anderson, *Lineages of the Absolutist State* (London: NLB, 1974).

12. This was true during the very early period. See Geoffrey Parker, "The Military Revolution,' 1560–1660—A Myth?," *Journal of Modern History* (1976): 195–214. It is still clearly true today; Britain's technological advantages over Argentina played a great role in her victory in the recent Falkland/Malvinas Islands war.

13. These political science jargon terms are discussed at length in Martin Wight *Power Politics* (New York: Holms & Meier, 1978), and in William P. Avery and David P. Rapkin (eds., *America in a Changing World Political Economy* (New York: Longman, 1982).

14. The dates, of course, are approximate. Different authors give somewhat different periods. It is clear, however, that the Dutch were the hegemonic power for much of the seventeenth century, the British for the late eighteenth and most of the nineteenth centuries, and the Americans for the mid-twentieth century.

15. On the threat of nuclear war see Jonathan Schell, *The Fate of the Earth* (New York: Knopf, 1982).

16. Indeed, if one takes a truly political-economic view of capitalism and the MWS, then it makes sense to say that an integrated world economy and international political anarchy (the state system) are two sides of the same coin. Each are defining characteristics of capitalism as a world system.

17. Wallerstein, "The Rise and Future Demise of the World Capitalist System," 404.

18. An instance of this just occurred at Arizona State University. In a ground-breaking ceremony for a new engineering college building, the first shovel was not dug by the president of the university, not by the dean of the college, not by the main donor to the building, but by an industrial robot using the same type of microprocessor technology with which this chapter was printed on an IBM Personal Computer!

19. A particularly outstanding example of such a book is David S. Landes, *The Unbound Prometheus: Technological Change and Industrial Development in Western Europe from 1750 to the Present* (Cambridge: Cambridge University Press, 1969).

20. The following discussion is based upon Immanuel Wallerstein, *The Modern World-System* vol. 2, 36–71, and David F. Noble, *America by Design: Science, Technology and the Rise of Corporate Capitalism* (New York: Knopf, 1977).

21. The Dutch, of course, built many other types of ships from small "buses" for the North Sea herring fishing fleet to giant "East Indiamen" for trade over very long distances in the Indian ocean and beyond. Ships of war were also built. Nevertheless, the basis of Dutch commercial supremacy was based upon trade in European waters using the ubiquitous flyboat.

22. Wallerstein, *The Modern World-System*, vol. 2, 43.

23. Ibid., 55.

24. Ibid., 46.
25. John Kenneth Galbraith, *The New Industrial State* (New York: Mentor Books, 3rd rev. ed., 1981).
26. Noble, *American by Design*, xxiv.
27. Karl Marx, *Grundrisses*. (Harmondsworth: Penguin Books, 1973).
28. The above discussion on the relationship between capital and the technology of transportation and communication is based upon David Harvey, "The Geography of Capitalist Accumulation: A Reconstruction of the Marxian Theory," in Richard Peet, ed., *Radical Geography: Alternative Perspectives on Contemporary Social Issues*, (Chicago: Maaroufa Press, 1977), 263–292. For an extensive discussion of current trends in communication technology see Rolf T. Wigand, "Direct Satellite Broadcasting: A Case for Appropriate Technology?" in this book.
29. Lane, *Profits from Power*.
30. Robert S. Cohen, "Science and Technology in Global Perspective," *International Social Science Journal* 34 (1982): 69.
31. Warren F. Ilchman and Norman T. Uphoff, "Beyond the Economics of Labor-intensive Development: Politics and Administration," *Public Policy* 22 (Spring 1974): 195. This is one of the few articles I have found that even begins to address the political issues involved in appropriate technological choices. It presents a good discussion of the administrative side of such decisions as well.
32. Ibid., 200–201. These preferences are also consistent with the alternative development strategy proposed by Diwan and Livingston and the basic human needs approach of Ghai. See Romesh Diwan and Dennis Livingston, *Development Strategies and Technological Choices in Developing Countries: Report Submitted to the National Science Foundation* (Troy), New York: Rensselaer Polytechnic Institute, 1978), and D. P. Ghai et al. *The Basic-Needs Approach to Development: Some Issues Regarding Technology and Methodology* (Geneva: International Labor Office, 1977).
33. Quoted in Schumacher, *Small Is Beautiful*, 153.
34. Partial exceptions are David Dickson, *The Politics of Alternative Technology* (New York: Universe Books, 1975); Denis Goulet, "The Suppliers and Purchasers of Technology: A Conflict of Interests," *International Development Review* 18 (1976): 14–20; and Ilchman and Uphoff, "Beyond the Economics of Labor-Intensive Development."
35. McRobie, *Small Is Possible*, 32.
36. These estimates are from Nicolas Jequier, ed., *Appropriate Technology: Problems and Promises* (Paris: Organization for Economic Cooperation and Development 1976), 36–37.
37. Michael P. Todaro, *Economic Development in the Third World* (New York: Longman, 2nd ed., 1981), 97.
38. This point is recognized by at least some advocates of appropriate technology.

But everywhere the administrative structure, the rules of the game, so to speak, favor the large against the small, the centralized operation against the decentralized, the rich against the poor. They bear the imprint of a system of technology and economics that puts goods before people. These rules relate to such matters as the availability and terms of finance, access to research and development facilities, management and technical training; freight rates and transport facilities; taxation policy; foreign exchange regulations; and economic planning, especially the criteria employed in assigning priorities to different kinds of economic activity.

McRobie, *Small Is Possible*, 187.
39. Cohen, "Science and Technology in Global Perspective," 62–63.
40. As discussed in McRobie, *Small Is Possible*, 183 and 188, and in Schumacher, 214.
41. McRobie, *op. cit.*
42. A point emphasized by Robertson, "Introduction: Technological Innovations," 438, and Cohen, "Science and Technology in Global Perspective," 69.
43. Certainly the impact of Schumacher's *Small Is Beautiful* illustrates what good ideas can do. He recognized the importance of a political ideology for appropriate technology but then immediately said, "it is not for me to talk politics." See 214 of *Small Is Beautiful*.
44. Ilchman and Uphoff, "Beyond the Economies of Labors Intensive Development," 196.
45. Schumacher, *Small Is Beautiful*, 240.
46. Todaro, *Economic Development in the Third World*, 94. This is one of the principal conclusions of the Nobel-Prize-winning research by Professor Simon Kuznets on the history of the modern economic growth of today's rich nations.

4. The Transfer of Technology to Third World Countries: Political Problems and International Ramifications

1. Harvey W. Wallender, III, *Technology Transfer and Management in the Developing Countries* (Cambridge, Mass.: Ballinger, 1979), p. 3.
2. See UNCTAD DOCUMENTS TD/B/C.6/AC/1/L.1/Rev. 1 (May 16, 1975) and TD/B/C.6/AC.1/L.6 (November 28, 1975) for Brazilian draft; and TD/B/C.6/AC.1/L.2 (May 5, 1975) and TD/B/C.6/AC/.1/L.5 (November 24, 1975) for the Japanese proposals.
3. Document TX/190 (December 31, 1975), p. 1.
4. Document TD/L.112 (May 27, 1976); TD/B/C.6/AC.3/2 (June 28, 1977); and TD/B/C.6/AC.3/3 (June 29, 1977).
5. Document TD/B/C.6/AC.1/L6, pp. 1–4.
6. For details see UNCTAD, Trade and Development Board, *Report of the Second Ad Hoc Group of Experts on Restrictive Business Practices October 20–24, 1975*, Document TD/B/C.2/AC.5/R.1 (November 10, 1975), p. 9.
7. U.S. Department of State, Bureau of Public Affairs, *Results of the Seventh Special Session of the United Nations General Assembly* September 1–6, 1975, p. 6.
8. U.S. Department of State, Bureau of Public Affairs, *UNCTAD IV: Expanding Cooperation for Global Development* (May 6, 1976), pp. 9, 10, 11.
9. U.S. Department of State, Bureau of Public Affairs, *Speech by Secretary Vance at the Northwestern Regional Conference on the Emerging International Order—March 30, 1979*, pp. 4–5.
10. *UN Draft International Code of Conduct on the Transfer of Technology as of 6 May 1981*. Document TD/Code TOT 25, pp. 4–5.
11. Ibid., pp. 6–7.
12. Ibid., pp. 7–8.
13. Ibid., pp. 9–10.
14. U.S. Department of State, Advisory Committee Memorandum, "Current Status of International Activities Relating to Transnational Enterprises (TNEs) as of December 1979," p. 9.
15. Ibid., pp. 8–9.
16. U.S. Department of State, Advisory Committee, Minutes of Meeting of Working Group on UN/OECD Investment Undertaking, January 23, 1979, p. 4.
17. U.S. Department of State, Advisory Committee, Minutes for Meeting of September 14, 1979, p. 13.
18. Draft International Code of Conduct on the Transfer of Technology, pp. 9–10.
19. Ibid., p. 9.
20. Ibid., pp. 11–19 give details of chaps. 4 and 5.
21. For details, see ibid. and UNCTAD Documents TD/Code TOT/27, November 17, 1980, p. 6.
22. For details, see Draft International Code of Conduct on the Transfer of Technology, pp. 6–7, and TD/Code TOT 25, pp. 15–19.
23. For details, see TD/Code TOT 25, pp. 20–23.
24. For details, see ibid., pp. 24–25.
25. See UN Document A/Conf. 81/16, 1979, esp. p. 110.
26. For a full background on these efforts, see UNCTAD Documents TD/B/C.6/AC.3/2 and 3/3 containing reports, respectively, on *The Revision of the Paris Convention for the Protection of Industrial Property* and *The Impact of Trademarks on the Development Process of Developing Countries*.
27. U.S. Department of State, "Current Status of International Activities Relating to Transnational Enterprises as of December 1979" (mimeographed), p. 12.
28. For further details, see UNCTAD Documents TD/Code TOT/27, pp. 10–11.
29. For details on the diverging positions, see ibid., pp. 13–14.
30. For a more comprehensive discussion of this issue, see ibid., pp. 15–18.
31. Ibid., pp. 18–20.
32. See Oren R. Young, "International Regimes: Problems of Conceptions Formations" *World Politics* (April 1980): 331–56.

5. From Dependency to Self-Reliance: An Evaluation of China's Experience of Technology Transfer

1. N. Rosenberg, "Technological Change in the Machine Tool Industry, 1840–1910," *Journal of Economic History* 23 (December 1963); N. Rosenberg, "The Direction of Technological Change: Inducement Mechanisms and Focussing Devices," *Economic Development and Cultural Change* 18 (October 1969); and Frances Stewart, *Technology and Underdevelopment* (Boulder, Colo.: Westview Press, 1977).

2. See, for example, A. G. Frank, *Capitalism and Underdevelopment in Latin America* (New York: Monthly Review Press, 1967); Theotonio dos Santos, "The Structure of Dependence," *American Economic Review* 40 (1970): papers and proceedings; C. Furtado, *Economic Development in Latin America* (Cambridge: Cambridge University Press, 1970); and Stewart, *Technology and Underdevelopment*.

3. The term 'suitability gap' was used by Paul Streen in *The Frontiers of Development Studies* (New York: John Wiley & Sons, 1972).

4. A. Emmanuel, *Unequal Exchange: A Study of the Imperialism of Trade* (New York: Monthly Review Press, 1972).

5. Hans Heyman, Jr., "Self-Reliance Revisited: China's Technology Dilemma," in Bryant G. Garth, ed., *China's Changing Role in the World-Economy* (New York: Praeger, 1975).

6. Jan S. Prybyla, *The Political Economy of Communist China* (Scranton: International Textbook Co., 1970), p. 100.

7. Li Fu-Chun, *Report on the First Five-Year-Plan for Development of the National Economy of the People's Republic of China in 1953–1957* (Peking: Foreign Language Press, 1955), p. 5–21.

8. Alexander Eckstein, *China's Economic Development* (Ann Arbor: University of Michigan Press, 1975), p. 33, table 4.

9. T. J. Hughes and D. E. T. Luard, *The Economic Development of Communist China, 1948–1958* (New York: Oxford University Press, 1959).

10. "On Questions of Party History," *Beijing Review*, no. 27 (July 6, 1981).

11. *Eleventh National Congress of the Communist Party of China*, document (Beijing: Foreign Language Press, 1977).

12. Rosali L. Tung, *Chinese Industrial Society After Mao* (Lexington, Mass.: D. C. Heath, 1982), pp. 35–48.

13. On the Dependency literature, see André Gunder Frank, *Latin America: Underdevelopment or Revolution* (New York: Monthly Review Press, 1969); A. G. Frank, *Capitalism and Underdevelopment in Latin America*, in James D. Cockcroft, A. G. Frank, and Dale L. Johnson, eds., *Dependence and Underdevelopment* (Garden City, N.Y.: Doubleday, 1972); A. G. Frank, *On Capitalist Underdevelopment* (New York: Oxford University Press, 1975); and Samir Amin, *Unequal Development* (New York: Monthly Review Press, 1976).

14. Susanne Bodeheimer, "Dependency and Imperialism: The Roots of Latin American Underdevelopment," *Politics and Society* 1, no. 3 (May 1971): 327.

15. Theotonio dos Santos, "The Structure of Dependence," *American Economic Review* 60 (May 1970).

16. R. Girling, "Dependency, Technology and Development," in F. Bonilla and R. Girling, eds., *Structures of Dependency* (Palo Alto: Nairobi, 1973).

17. Cited in Dennis M. Ray, "Chinese Perceptions of Social Imperialism and Economic Dependency: The Impact of Soviet Aid," in Bryant G. Garth, ed., *China's Changing Role in the World Economy* (New York: Praeger Publishers, 1975).

18. Prybyla, *Political Economy of Communist China*, p. 218.

19. Subramanian Swamy, *Economic Growth in China and India, 1952–1970* (Chicago: University of Chicago Press, 1973), p. 82.

20. Sulslov, "On the Struggle of the C.P.S.U. for the Solidarity of International Communist Movement," *Current Digest of the Soviet Press* (April 29, 1964).

21. Cited in Alfred D. Low, *The Sino-Soviet Dispute: An Analysis of the Polemics* (Rutherford, N.J.: Fairleigh Dickinson University Press, 1976), pp. 57–58.

22. "Letter of the Central Committee of the CPC of February 29, 1964 to the Central Committee of

the CPSU," *Peking Review* (May 8, 1964), p. 13; also cited in Prybyla, *Political Economy of Communist China*, p. 218; and Ray, "Chinese Perceptions of Social Imperialism," p. 61.

23. Benjamin Schwartz, "Sino-Soviet Relations—The Question of Authority," *The Annals of the American Society* 349 (September 1963).

24. Johan Galtung, "A Structural Theory of Imperialism," *Journal of Peace Research* 8 (1971): 81–117.

25. China's import values and GNP from 1950 to 1976 are available in *China—Facts & Figures Annual* (Gulf Breeze, Fla.: Academic International Press, 1978).

26. The figures indicating the direction of Chinese trade from 1950 to 1964 are available in Feng-Hwa Mah, "Foreign Trade," in Alexander Eckstein et al., eds., *Economic Trends in Communist China* (Chicago: Aldine, 1968).

27. For China's trade balance figures, see table 1.

28. Average commodity composition of Chinese foreign trade during 1967–75 is as follows: Export manufactures 42 percent, foodstuffs 30 percent, crude materials 21 percent, chemicals 5 percent, other 2 percent; Import manufactures 51 percent, foodstuffs 18 percent, crude material 17 percent, chemicals 13 percent, other 1 percent. Cited in Bohdan O. and Maria R. Szaprwicz, *Doing Business with the Peoples Republic of China* (New York: John Wiley & Sons, 1978).

29. Hans Heymann, Jr., "Self-Reliance Revisited: China's Technology Dilemma," in Garth, ed., *China's Changing Role in the World Economy*, pp. 26–28.

30. Thomas G. Rawski, "Choice of Technology and Technological Innovation in China's Economic Development," in Robert F. Derberger, ed., *China's Development Experience in Comparative Perspective* (Cambridge, Mass.: Harvard University Press, 1980).

31. Ibid.

6. Nuclear Power in the Philippines: Technological Choice and Dependence

1. An earlier version of this paper was presented at the Annual Meeting of the Western Political Science Association, March 25–27, 1982, Bahia Hotel, San Diego.

2. "The Philippines Ponder What to Do About Nuclear Power," *Nucleus* 2 (Summer 1980): 4.

3. Johan Galtung, "A Structural Theory of Imperialism," *Journal of Peace Research* 8 (1971): 83.

4. Walden Bello, Peter Hayes, and Lyuba Zarsky, "'500-Mile Island': The Philippine Nuclear Reactor Deal," *Pacific Research* 10 (first quarter, 1979):15.

5. *Bulletin Today* (Manila), March 2, 1981, p. 24.

6. Bello et al., "'500-Mile Island,'" p. 9.

7. Ibid., pp. 7–10. As of March 1981 all of the suits were settled except for one with the Long Island Lighting Company, a would-be recipient of seven million pounds of uranium. See Barnaby J. Feder, "Is Westinghouse Out of the Woods?," *New York Times*, March 15, 1981, sec. 3, pp. 1, 22.

8. *Ang Katipunaan*, March 1–15, 1978, reprinted in *Philippine Liberation Courier* (Oakland, Calif.), March 10, 1978, pp. 2–3. See also U.S. Congress, House of Representatives, Subcommittee on Foreign Operations Appropriations, *Foreign Assistance and Related Agencies Appropriations for 1979, Part 1*, 95th Cong., 2nd sess., February 8, 1978, pp. 97–125.

9. Barry Kramer, "Ties to the Top," *Wall Street Journal*, January 12, 1978, pp. 1, 18. Disini's wife, Inday Escolin, is a first cousin and a physician of Imelda Marcos, the president's wife and a former governess of the Marcos's three children. See *Philippine Times* (Chicago), December 16–January 31, 1978, p. 15, and February 1–28, 1978, p. 11.

10. Fox Butterfield, "Marcos, Facing Criticism, May End $1 Billion Westinghouse Contract," *New York Times*, January 20, 1978, p. A3; and "Uranium and the Third World," a Friends of the Earth Broadsheet (1977), p. 2.

11. *Ichthys* (Manila) April 7, 1978, pp. 1–2, and *Philippine Liberation Courier*, May 12, 1978, p. 2.

12. *Bulletin Today*, July 1, 1979, pp. 1, 10; *Daily Express*, (Manila), July 18, 1979, p. 1; and *Bulletin Today*, October 4, 1979, p. 28.

13. Romeo Pajarillo, "Nuclear Plant Critics Walk Out as Hearing Ends," *Bulletin Today*, October 4, 1979, p. 1.

14. *Bulletin Today*, August 20, 1979, pp. 1, 10 and August 21, 1979, pp. 1, 5.

15. Lorenzo M. Tañada, "In Constructing the Bataan Nuclear Plant, The Government Is More Accommodating to Westinghouse, A Foreign Concern Than Its Citizenry," *Philippine Panorama* (Manila), September 13, 1981, p. 16.

16. Thomas O'Toole, "Officials Scored on A-Plant Loan," *Washintgon Post*, February 9, 1978, p. A21, and *Philippine Times*, February 10–28, 1978, pp. 3, 11, 22.

17. Daniel F. Ford to President Ferdinand E. Marcos, September 13, 1978, Union of Concerned Scientists (Cambridge, Mass.), and *Bulletin Today*, August 10, 1979, pp. 1, 5.

18. Dennis M. O'Leary, "Westinghouse Hit at Rally," *Philippine Times*, April 27–May 3, 1978, p. 14, and *Philippine Liberation Courier*, April 14, 1978, pp. 1, 8.

19. Brian Firth, "Green Light for Yellowcake," *Far East Economic Review*, November 17, 1978, pp. 84, 87; *Asiaweek*, September 22, 1978, pp. 35–36; and David Solomon, "Aboriginal Rights in Australia Meeting the Nuclear Age," *Christian Science Monitor*, September 28, 1978.

20. Joanne Omang, "A-Panel Allows Reactor Export to Philippines," *Washington Post*, May 7, 1980, p. A17, and *Bulletin Today*, August 9, 1979, pp. 1, 11.

21. Congressman Clarence D. Long to NRC Chairman Joseph Hendrie, January 4, 1978 (Washington, D.C.), and *Bulletin Today*, July 5, 1979, pp. 1, 8.

22. Address of Robert D. Pollard before the Rotary Club of Manila, March 19, 1981, reprinted in *NASSA News*, (Pasay City, Philippines), March–April 1981, p. 8.

23. *Arizona Republic*, April 26, 1981, p. A5, and Matthew L. Wald, "Steel Turned Brittle by Radiation Called a Peril at 13 Nuclear Plants," *New York Times*, September 27, 1981, pp. A1, A64.

24. *Arizona Republic*, February 2, 1982, p. A1, and February 13, 1982, p. A22

25. *Arizona Republic*, January 26, 1982, p. A12, and February 11, 1982, p. A15.

26. Pollard, *NASSA News*, March–April, 1981, pp. 7–11, and Michael Richardson, "Manila's Disillusion with the Atom," *Far Eastern Economic Review*, June 23, 1978, pp. 94, 96–97, 99.

27. Tañada, "Bataan Nuclear Plant," p. 50, and the letter of Long to Hendrie, January 4, 1978.

28. *Daily Express*, July 21, 1979, pp. 1, 7; O'Toole, "Officials Scored on A-Plant Loan," p. A21; and Monica S. Feria, "Nuclear Plant Sits on Volcanic Site," *Daily Express*, July 7, 1979, pp. 1, 6.

29. *Bulletin Today*, August 16, 1979, pp. 1, 8.

30. *Philippine Times*, February 10–28, 1978, p. 11.

31. Paul Icamina, "N-Plant Rising in Bataan," *Daily Express*, April 29, 1978, pp. 1, 13.

32. *Ichthys*, July 7, 1978, pp. 12–17.

33. I. L. Urmeneta, "Power for Whom?", Community Organization Research and Document Program, Development Issues Series No. 2. (University of the Philippines: Institute of Social Work and Community Development, n.d.), p. 6, and Henry Kamm, "Islanders Fight Japan's Plan to Dump Atom Waste," *New York Times*, March 18, 1981, p. A7.

34. Ralph Blumenthal, "Jersey Reactor Creates Tax Gains and Radioactive Waste Problem," *New York Times*, March 18, 1981, p. A7.

35. *Philippine Liberation Courier*, December 16, 1977, p. 6.

36. Bello et al., "500-Mile Island," p. 15.

37. Denis Hayes, "Nuclear Economics," in Mark Reader, Ronald A. Hardert, and Gerald L. Moulton, eds., *Atom's Eve: Ending the Nuclear Age* (New York: McGraw-Hill, 1980), p. 79, and Urmeneta, "Power for Whom?", p. 5.

38. Bello et al., "500-Mile Island," pp. 27–28.

39. *Signs of the Times* (Manila), September 11, 1976, pp. 15–17.

40. *Tanod* (Oakland, Calif.), September 1978, pp. 5–6.

41. Manila Domestic Service, March 28, 1978, in United States Foreign Broadcast Information Service (FBIS), *Daily Reports* 4 (Asia and Pacific, March 29, 1978): P1.

42. *Bulletin Today*, July 24, 1979, pp. 1, 8, and Geronimo G. Velasco, *Development, Energy, and National Survival* (Makati: Ministry of Energy, 1979), p. 52.

43. *Times Journal* (Manila), February 21, 1978, p. 1.

44. Alberto Rous, "Bataan Nuclear Plant Safe," *Times Journal*, April 29, 1978, p. 1.

45. Tony Antonio, "Allay Nuclear Plant Fears," *Bulletin Today*, July 6, 1979, pp. 1, 8.

46. Philippines (Republic), *Long Term Philippine Development Plan Up to the Year 2000* (Manila: Department of Economic Planning, 1977), p. xviii.

47. Bello et al., "500-Mile Island," p. 16, and Christopher S. Gray, "The Philippines—Foreign Investment Surges," *Business Week*, June 29, 1981, pp. 20–40 (special advertisement section).

48. For critical analyses of foreign investment in the Philippines, consult Corporate Information Center for the National Council of Churches of Christ in the U.S.A., *The Philippines: American Corporations, Martial Law, and Underdevelopment* (New York: IDOC–North America, Inc., 1973); Augusto C. Espiritu et al., *Philippine Perspectives on MultiNational Corporations* (Quezon City: University of the Philippines Law Center, 1978); and John F. Doherty, ed., *Readings on Peripheral Development: The Role of the Multinationals* (Manila: n.p., 1978).

49. Bello et al., "500-Mile Island," pp. 16–17.
50. "Summary of the Findings of a World Bank–Asian Development Bank Survey of the Philippine Energy Situation," *Business Day* (Manila), March 4, 1981, p. 20.
51. Philippines (Republic), National Economic and Development Authority (NEDA), *1978 Philippine Statistical Yearbook* (Manila: NEDA, 1978), p. 71, table 1.18.
52. *Daily Express*, October 24, 1979, p. 6.
53. Judith Tendler, "Rural Infrastructure Projects: Roads and Electrification," paper prepared for the Bureau of Program and Policy Coordination, USAID, October, 1978, p. 128, as quoted in Bello et al., "500-Mile Island," p. 19.
54. See Lindy Washburn, "Our Lake for Others; The Maranao and the Agus River Hydroelectric Project," and Gene Stoltzfus and Dorothy Friesen, "Erosion and Soil Depletion—By-Products of Castle and Cooke Operations," *MSPC Communications* (Davao City, Philippines), no. 27 (April 1978), pp. 11–14 and pp. 17–19; Noel Yamada, "The Poisoning of Cagayan de Oro: Kawasaki Steel in the Philippines," *AMPO Japan-Asia Quarterly Review* 12, no. 2 (1980): 70–73; and "Tribal People and the Marcos Regime," *Southeast Asia Chronicle*, no. 67 (October 1979), pp. 1–32.
55. David Dickson, *The Politics of Alternative Technology* (New York: Universe Books, 1974), p. 183.
56. Ibid., pp. 59–60.
57. *Business Week*, June 12, 1978, p. 26.
58. André Gunder Frank, "Sociology of Development and Underdevelopment of Sociology" in James D. Cockcroft, André Gunder Frank, and Dale L. Johnson, eds., *Dependence and Underdevelopment: Latin America's Political Economy* (Garden City, N.Y.: Doubleday, Anchor Books, 1972), pp. 365–66.

7. Intermediate Technology in Newly Industrialized Countries: Two Cases from South Korea

1. Parvey Hasan and D. C. Rao, *Korea: Policy Issues for Long-Term Development* (Baltimore: The Johns Hopkins University Press [for the World Bank], 1979), pp. 3–4.
2. U.S. Embassy, Seoul, "Economic Trends Report," January 12, 1979, pp. 4–5.
3. *The Hankook Ilbo*, December 19, 1979, p.1.

8. Weapons Production in Less Developed Countries: A Possibility for Integrating Technology with Development?

1. Mary Kaldor, "The Military in Development," *World Development* 4, no. 6 (1976): 476.
2. Reference to development in LDCs means the improvement of economic, social, *and* political conditions within those countries.
3. Richard Falk, "Militarization and Human Rights in the Third World," in A. Eide and M. Thee, eds., *The Problems of Contemporary Militarism* (New York: St. Martin's, 1980), pp. 220–25.
4. This chapter uses the terms "arms" and "weapons" interchangeably to include ammunition, support equipment, and spare parts.
5. International Peace Research Association (IPRA), "The Impact of Militarization on Development and Human Rights," *Bulletin of Peace Proposals* 9, no. 2 (1978): 1976.
6. M. Wolpin, "Arms Trade and Transfer of Military Technology," in Eide and Thee, eds., *Problems of Contemporary Militarism*, p. 245.
7. Jan Oberg, "Arms Trade with the Third World as an Aspect of Imperialism," *Journal of Peace Research* 12, no. 3 (1975): 213.
8. I. Peleg, "Military Production in Third World Countries: A Political Study," in P. McGowan and C. W. Kegley, Jr., eds., *Threats, Weapons and Foreign Policy* (Beverly Hills: Sage, 1980), pp. 209–30.
9. IPRA, "Impact of Militarization," p. 175.
10. Ibid.
11. U. Albrecht et al., "Militarization, Arms Transfer and Arms Production in Peripheral Countries," *Journal of Peace Research* 12, no. 3 (1975): 206.
12. P. Lock and H. Wulf, "Consequences of Transfer of Military Oriented Technology on the Development Process," *Bulletin of Peace Proposals* 7 (1977): 130.

13. Dieter Ernst, "International Transfer of Technology, Technological Dependence, and Development Strategies," *Bulletin of Peace Proposals* 10, no. 2 (1979):202.
14. Ibid.
15. Lock and Wulf, "Consequences of Transfer of Military Oriented Technology," p. 132.
16. Jan Oberg, "The New International Economic and Military Orders as Problems to Peace Research," *Bulletin of Peace Proposals* 8 (1977):143; Kaldor, "Military in Development," p. 476.
17. Emile Benoit, *Defense and Economic Growth in Developing Countries* (Lexington, Mass.: D.C. Heath, 1973), pp. 2–3.
18. See chapters in Henry Bienen, ed., *The Military and Modernization*. Chicago: Aldine and Atherton, 1971; or J. J. Johnson, *The Role of the Military in Underdeveloped Countries* (Princeton: Rand Corporation, 1962).
19. Oberg, "Arms Trade with the Third World," p. 146.
20. H. Kitamura, "Rationale and Relevance of the New International Economic Order," *The Developing Economics* 16, no. 4 (1978):343.
21. See L. Kraar, "Israel's Own Military-Industrial Complex," *Fortune*, March 13, 1978, pp. 72–76, for a description of Israel's export and domestic arms production successes.
22. "Brazil: Arms to Libya," *Latin America Political Report* 4 (March 1977): 66–68.
23. U. Albrecht and Mary Kaldor, "Introduction," in Kaldor and A. Eide, eds., *The World Military Order* (New York: Praeger, 1979), p. 8.
24. R. Vayrynen, "The Arab Organization of Industrialization: A Case Study in the Multinational Production of Arms," *Current Research on Peace and Violence* (1979):66–79.
25. R. Ropelewski, "Arabs Push Arms Industry Despite Peace Expectations," *Aviation Week and Space Technology*, November 16, 1977, p. 16.
26. "Arabs Expected to Map Joint Military Strategy," *Arizona Republic*, March 16, 1981, p. A14.
27. Benoit, *Defense and Economic Growth*, pp. 213–18.
28. "U.S. Weapons Deal with Pakistan Could Start Arms Race, India Says," *Arizona Republic* April 19, 1981, p. A14.
29. Signe Landgren-Backstrom, "The Transfer of Military Technology to Third World Countries," *Bulletin of Peace Proposals* 7 (1977):120.

9. Direct Satellite Broadcasting: A Case for Appropriate Technology?

This research was supported by a grant from the International Social Science Council, Paris. The author gratefully acknowledges the assistance received from scholars and organizations in over twenty countries, too numerous to recognize individually.

1. R. E. Rice and E. B. Parker, "Cost and Policy Implications of Communication Satellite Systems for Rural Telecommunications in Developing Countries," paper presented at the seventh annual Telecommunications Policy Research Conference, Skytop, Pa., April 29–May 1, 1979; Rolf T. Wigand, "The Direct Satellite Connection: Definitions and Prospects," *Journal of Communication* 30 (1980): 140–46.
2. A. T. Henderson, National Communications Satellite Evaluation Support Group, Australian Postal and Telecommunications Department, personal communication, August 6, 1979; J. Grey and M. Gerard, *A Critical Review of the State of Foreign Space Technology* (New York: American Institute of Aeronautics and Astronautics, 1978), pp. 99–124; "The Global Communications Satellite Catalogue," *Satellite Communication* (January 1978): 27–30.
3. S. A. Kader, secretary general, Arab States Broadcasting Union, personal communication, July 7, 1979.
4. Wigand, "Direct Satellite Connection"; Rolf T. Wigand, "Satellitenkommunikation: Die heutige Situation in Nordamerica," *Media Perspektiven* 7 (1980): 445–56.
5. Rolf T. Wigand, "The Future of Telecommunications," paper presented at the Telecommunications in the Year 2000: National and International Perspectives Conference, New Brunswick, N.J., November 17–20, 1981; Rolf T. Wigand, "The Direct Satellite Connection: Its Future and Policy Implications," paper presented at the annual convention of the International Communication Association, Acapulco, Mexico, May 21, 1980.
6. United Nations, "Report of the Legal Sub-Committee on the Work of Its Eighteenth Session (12 March–6 April 1979)," A/AC. 105/240, April 10, 1979.
7. M. A. Dauses, "Direct Satellite Broadcasting by Satellites and Freedom of Information," *Journal of Space Law* 3 (1975): 59–72.

8. A. W. Frutkin, "Direct/Community Broadcast Projects Using Space Satellites," *Journal of Space Law* 3 (1975): 17–24; E. Galloway, "Direct Broadcast Satellites and Space Law," *Journal of Space Law* 3 (1975): 3–16; I. M. Pikus, "Legal Implications of Direct Broadcasting Technology," *Journal of Space Law* 3 (1975): 17–24.

9. Wigand, "Direct Satellite Connection."

10. International Telecommunication Union, *Final Protocol: Space Telecommunications*, [1972] 23 U.S.T. at 1573, n. 1, July 17, 1971 (effective January 1, 1973).

11. Wigand, "Direct Satellite Connection."

12. D. A. Golden, personal communication, July 3, 1979.

13. Ibid.

14. Wigand, "Direct Satellite Connection."

15. "Business Communications: The New Frontier, *Fortune*, October 9, 1978, pp. 26–80; P. L. Bargellini, "A Synopsis of Commercial Satellite Communication Systems," unpublished paper, Comsat Laboratories, 1977; N. Engler, J. Strange, and G. Hein, *Compendium of Applications Technology Satellite User Experiments 1967–1973* (Cleveland: NASA-Lewis, August 1976).

16. Rolf T. Wigand, "Local Communication and Communication Research: An American Perspective," paper presented at the Congress on Local Communication and Communication Research, International Union for Research on Communication, Baden-Baden, November 12–13, 1981. Also published in *Communications—International Journal for Communication Research* 9 (1982): 189–217.

17. J. Sauve, "Communications Satellites: The Canadian Experience," unpublished paper, Montreal, 1978.

18. J. H. Disher, "Space Transportation, Satellite Services, and Space Platforms," *Astronautics and Aeronautics* 17 (1979): 42–51, 67.

19. Ibid.; F. Press, "U.S. Space Policy—A Framework for the 1980s," *Astronautics and Aeronautics* 17 (1979): 34–35; C. J. Goodwin, "Space Platforms for Building Large Space Structures," *Astronautics and Aeronautics* 16 (1978): 44–47; B. J. Edelson and W. L. Morgan, "Orbital Antenna Farms," *Astronautics and Aeronautics* 15 (1977): 20–28.

20. T. A. Eastland, "Satellite Communications and Related Applications," in Department of Communications, *Research and Development* (Ottawa: Government of Canada, 1977).

21. J. P. Taylor, "Satellite Systems of the 1990s Will Operate from Huge Platforms Orbiting in Space," *Television/Radio Age* 25 (1978): 25–29, 124–26.

22. Edelson and Morgan, "Orbital Antenna Farms."

23. Goodwin, "Space Platforms for Building Large Space Structures."

24. Edelson and Morgan, "Orbital Antenna Farms."

25. I. Bekey, "Big Comsats for Big Jobs at Lower User Cost," *Astronautics and Aeronautics* 17 (1979): 42–56.

26. Wigand, "Direct Satellite Connection"; Wigand, "Future of Telecommunications."

27. Bekey, "Big Comsats."

28. Taylor, "Satellite Systems of the 1990s."

29. Rolf T. Wigand, "Mass Media Abundance: Selected Developments and Audience Effects in the United States of America," *Communications—International Journal of Communication Research* 5 (1979): 2–3, 213–39.

30. Rolf T. Wigand, "The Process of Mass Communication: Mass Media as Agents of Socialization in Developing Countries," *Konferenzprotokoll der Association Internationale des Etudes et Recherches sur l'Information* 2 (1976): 473–86.

31. International Telecommunications Union, *Final Acts: WARC 1979* (Geneva: ITU, 1979).

32. Wigand, "Process of Mass Communication."

Index

Abundant Harvest Pipehouse Co., 102–4, 107, 108
Aetna Life and Casualty, 123
Africa: mini-brickworks in, 26; satellite systems in, 123, 126, 130; textile production in, 11
Afrosat satellite system, 130
Agriculture: in Korea, 102–4; no-till system of, 25; in Philippines, 95–96
Algeria: armaments in, 117–18; satellite systems in, 122, 126
Alternative development, strategies of, 148
American Bar Association, 127
American College of Physicians and Surgeons, 127
American Federation of Scientists, 88
American Hospital Association, 127
American Library Association, 127
American Medical Association, 127
American Telephone and Telegraph (AT&T), 122, 132
Andean Common Market, 63
Andean nations, satellite systems in, 122, 123, 126
Anik satellite system, 132
Appalachian region, 31
Applications Technology Satellites (ATS), 134
Appropriate technology, definition of, 3–5
Appropriate Technology (journal), 19
Appropriate Technology Development Association, 23, 26
Appropriate Technology for the United Kingdom (AT-UK), 28
Arab nations, satellite systems in, 130
Arab Organization of Industrialization (AOI), 116
Arabsat satellite system, 130
Argentina, arms production in, 113
Arms expenditures, of developing nations, 110–14
Arroyo, Joker, 87–88
Arms transfer (AT), 109–19 passim
Asia Industries Incorporated, 86
Asian Development Bank, 95
ATS-1 satellite system, 132
Aussat satellite system, 132
Australia, satellite systems in, 123, 132

Bangladesh Planning Commission, 24
Bataan Export Processing Zone (BEPZ), 94
Bataan Nuclear Power Plant. *See* Philippine Nuclear Power Plant

Bekey, I., 137–38
Belgium, satellite systems in, 129
Bell Trade Act of 1946, 84
Benoit, Emile, 117–18; *Defense and Economic Growth in Developing Countries*, 114–15
Bienen, Henry, 109
Braudel, Fernand, 32
Brazil: armaments in, 110, 111, 113; arms exports of, 115–16; auto industry in, 36; and code of conduct for transfer of technology, 51; satellite systems in, 122, 123, 126, 130
Brazilsat satellite system, 130
Britain. *See* Great Britain
British Coal Board, 16
British Hildon Trust, 21

Canada, satellite systems in, 127–28, 129, 132
Canadian Broadcasting Corporation, 135
Canadian Department of Communications, 128, 129, 134
Capital-intensive technology, 38; biases toward, 6–7; dynamics of, 34–36; limitations of, in developing nations, 16
Capitalist commodity production, 34–36
CBSS satellite system, 132
Center nations, 83–98 passim. *See also* Core economies
Charter of Rights and Duties of States, 49
Chile, satellite systems in, 123, 126
China: import dependence in, 77; mini-cement plants in, 26; satellite systems in, 122, 126, 131; technology transfer experience of, x, 67–81 passim
Chinese Communist Party, 72
Choi Kyu Hah, 102
Chun Doo Hwan, General, 101
Civil Action Military Assistance Program (CAMAP), 117
Coalition for a Nuclear-Free Philippines, 89
Code of conduct on transfer of technology, 49–63 passim
Cohen, Robert S., 44
Colgate Palmolive Philippines, 94
Colombia, satellite systems in, 123, 124, 130
Committee Against Nuclear Pollution in the Philippines, 89
Communications satellites. *See* Direct satellite broadcasting (DBS)
Comstar satellite system, 132
Confucianism, 99
Consat General Corporation, 123

Consolidated Edison Indian Point 2 nuclear plant, 90
Core economies, 31–47 passim
Cultural Revolution, Chinese, 69, 72, 80

Dauses, M. A., 124
David Livingston Institute of Overseas Development Studies, 11
Davis, John, 28
Denmark, satellite systems in, 126, 129
Départements satellites, 133
Development Techniques Ltd., 21
Direct satellite broadcasting (DBS), x, 44, 121–43 passim
Disini, Herminio, 86–87
Domestic arms production (DAP), 109–19 passim

Ebasco Services, Inc., 90
Eckaus, Richard S., 3, 5
Economic Commission for Asia and the Pacific, 24
Ecuador, satellite systems in, 123
Egypt: satellite systems in, 126; weapons production in, 116
Ekran satellite system, 132
Energy Development Board, 91
Energy Resource Group, Inc., 93
Engineering profession, in nineteenth-century America, 41–42
Ernst, Dieter, 114
Ethiopia: footwear manufacturing in, 11; windpower in, 27
European Communications Satellite (ECS), 132
European Space Agency (ESA), 127, 129, 133
Eutelsat satellite system, 126, 132
Export-Import Bank (Eximbank), 85, 86, 88

Falk, Richard, 109
Falkland Islands crisis, 115, 147
Federal Republic of Germany. *See* Germany
Ferg, David, 93
Feudalism, European, 33, 36
First Five-Year Plan, Chinese, 69, 70, 74
Fishing industry, in Philippines, 91, 95
Flyboat, sixteenth-century Dutch, 39–40, 44, 148
Forbes, Ian, 93
Ford, Daniel, 88
Four Modernizations, Chinese, 69, 72–73, 78, 80
France: satellite systems in, 126, 128, 129, 132–33; state-capital merger in, 38
Frank, André Gunder, 97
Frutkin, A. W., 124

Galloway, E., 124
Galtung, Johan, 76, 83, 84

Gandhi, Mahatma, 5, 45
General Electric Co., 85, 92; GE-UK, 35
General Telephone Co. (GTE), 132
Germany, satellite systems in, 123, 126, 129, 132–33
Gorizont satellite system, 132
Great Britain: intermediate technology programs, 19–23, 27, 28; satellites in, 129
Great Crisis and Readjustment, Chinese, 69, 72
Great Leap Forward, Chinese, 69, 70, 71–72, 75
Greenhouses, in Korea and Japan, 102–3
Group of 77, 59–62
Gulf Corporation Council, 117

Halcrow, William, 27
Hamburg Group, 115
Hegemony, 37–40, 147
Hendrie, Joseph, 88
Herdis Management and Investment Corporation, 86
Hippel, Frank von, 88
Holland. *See* Netherlands

Imperialism, and arms technology, 115
India: arms production in, 111–13; intermediate technology in, 5, 22, 26; satellite systems in, 122, 126, 130
Indigenous technology, 53
Indo-Pakistani War, 116
Indonesia, satellite systems in, 122, 131
Inmarsat satellite system, 127
Insat satellite system, 130
Intelsat satellite system, 122–23, 125–26, 128, 132
Intermediate technology, 5; attitudes toward, 46; definition of, 146; in developed countries, 27–28; in developing countries, 15–29 passim; and military production, 113–14, 118–19; in South Korea, 99–108
Intermediate Technology Development Group, x, 20–29; catalogue of, 19; founding of, 19, 45
International Business Machines (IBM), 123
International Labor Office (ILO), 12, 23, 24, 25
International Peace Research Association (IPRA), 110, 112
International Telecommunications Consortium, 125
Intersputnik satellite system, 132
Iran, satellite systems in, 122
Iraq, satellite systems in, 126
Ireland, satellite systems in, 129
Iscom satellite system, 130
Israel, armaments in, 110–11
Italy, satellite systems in, 129
IT Transport Ltd., 24

Jamaica, 20

Index

Japan: colonial legacy of, in Korea, 99, 107; nuclear power in, 89, 91; satellite systems in, 122; state-capital merger in, 38; technological influence of, on South Korea, 98–108; as technology-source country, x, 108

Kaddafi, Colonel Moammar, 115
Kaldor, Mary, 109, 114
Kennedy, Gavin, *The Military and the Third World*, 114–15
Kenya: intermediate technology in, 9, 10, 12, 23, 25, 27; Rural Access Roads Program in, 23
Kilusang Kabuhayan at Kaunlaran (KKK), 84
Kim Dae Jung, 102
Kim Il-sung, 99
Kissinger, Henry, 53–54
Korea, North, 99
Korea, South, 94; intermediate technology in, 99–108; technology transfer in, x
Korean War, 69, 99

Labor, hierarchical division of, 32–33
Labor-intensive technology, 4–14 passim
Landgren-Backstrom, Signe, 119
Landsat satellites, 127
Lane, Frederick, 35
Leeds, David, 90
Libya, arms sales to, 115–16
Long, Clarence, U.S. Congressman, 88
Loutch satellite system, 132

McBain, M. S., 11
McCulloch, Rachel, 61
McRobie, George, *Small Is Possible*, 28
Malaysia, satellite systems in, 126
Manchuria, 75–76
Mao Zedong, 72
Marcos, President Ferdinand, 84–97 passim
Marecs satellite system, 132
Marinduque Mining and Industrial Corporation, 94
Marisat satellite system, 127, 132
Marx, Karl, 32, 43–44
Messerschmitt-Bolkow-Blohm (MBB), 133
Mexico, satellite systems, 131
Middle East, satellite systems in, 122
Miles, Derek, 25
Military Bases Agreement (1947), 84
Military technology. *See* Weapons production
Molniya satellite system, 132

NASA, 122, 123, 126, 134–35
National Academy of Sciences, 3, 12
National Agriculture Library, 53
National Rural Electrification Cooperative Association (U.S.), 95
Nazareno, Ernesto, 92
Neo-mercantilism, 38

Netherlands, the: satellite systems in, 129; shipbuilding in, 36, 39–41, 44, 148
New International Economic Order (NIEO), 45, 47, 49, 62, 115
Nigeria: ITDG work in, 20; palm-wine distilleries in, 9; satellite systems in, 122, 126
Nobles, David, 41, 42
Norway, satellite systems in, 122, 126
Nuclear power: in Japan, 88, 91; in U.S., 87, 89, 90; in Philippine Islands, 83–98 passim
Nuclear Regulatory Commission (NRC), 89

Oberg, Jan, 111; and "armament imperialism," 114, 115
OECD Guidelines for Multinational Corporations, 56
Overseas Development Administration, 21, 23

Palapa satellites, 126, 131
Paris Convention for the Protection of Industrial Property of 1883, 51, 62; U.S. position on, 58
Park Chung-hee, President, 100, 101
Pedal power, 10
Periphery nations, 83–97 passim
Peru, satellite systems in, 126, 133
Philippine Islands: nuclear power in, x, 83–97 passim; satellite systems in, 123
Philippine Nuclear Power Plant (Bataan): decision to build, 85; commission on safety of, 88; controversies regarding, 89–92; implications of, for future of Philippines, 93–98
Pickett, J., 11
Pikus, I. M., 124
Pipehouses, Korean, 102–4
Pollard, Robert, 89
Power Contractors, Inc., 86, 90
P.R.C. satellite system, 131
Preparatory Intergovernmental Committee (PIC), 58
Proctor and Gamble Philippines, 94
Public Service Satellite Consortium, 127
Public works, rural, 11–13
Puno, Ricardo, 88; Puno Commission, 88, 93

Qatar, armaments in, 116

Radio broadcasting. *See* Direct satellite broadcasting
Raduga satellite system, 132
RCA Satcome satellite system, 132
Reading University, and ITDG, 21
Reagan, President Ronald, 102
Reconstruction period, Chinese, 69, 70
Renault auto firm, 35, 54
Renewable energy resources, 26–27
Restrictive business practices (RBPs), 52, 53, 60;

160 Appropriate Technology

Restrictive business practices *(continued)*
   list of, in code of conduct for transfer of technology, 56–57
Rodrigo, Francisco "Soc", 87
Rural Access Roads Program (Kenya), 12

Satcol satellite system, 130
Satellite Business Systems (SBS), 123
Satellite communication systems: corporate uses of, 138; international, 132–33; national, 122–23, 130–31; personal uses of, 137–38
Satellite Television Corporation, 132
Satmex satellite system, 131
Saudi Arabia: armaments in, 110, 116; satellite systems in, 122–23
Scandinavia, satellite systems in, 122, 126
Schumacher, E. F., x, 5, 15, 28, 31, 32, 34, 35; adviser to British Coal Board, 16; and founding of Intermediate Technology Development Group (ITDG), 19, 45; on India, 18–19; optimism of, 46–47; *Small Is Beautiful*, 21, 44; vision of, 146
Shipbuilding industry, seventeenth-century Dutch, 33, 39–41, 147
Sino-Soviet trade, 71
Smith, Adam: pin factory plan, 42; *The Wealth of Nations*, 35
Soft technology, 46–47
Solar energy, 21, 27
South Africa, armaments in, 111, 113
Soviet Union, 122, 123; and arms sales, 118; economic and technical aid of, to China, 67–81; satellite systems in, 128, 132–33
Space, satellite uses in, 135–36
Space shuttle, 135–36
Spain, satellite systems in, 126, 129
Sputnik satellite system, 122
Stationar satellite system, 132
Stewart, Francis, 9
STW satellite system, 131
Subic Bay Naval Base, 84, 90
Sudan: Rahad project in, 7; river current turbines in, 21, 27; road-building program in, 12; satellite systems in, 122–23, 126
Summa Insurance Corporation, 87
Sunil Electronics Co., 105–8
Sweden, satellite systems in, 129, 133
Switzerland, satellite systems in, 129
Symphonie satellites, 126, 132–33

Tañada, Lorenzo, 87–88
Tanzania, intermediate technology in, 20, 26
Technology dependence, definition of, 73–74
Technologies: advanced, 147; current, 8; indigenous, 8, 11; older, 8
Technology: implantation of, 4; substitution, 4
Technosphere Consultants Group, 86–87

Telecom satellite system, 133
Telephone services, 127
Telesat, 135; Telesat Canada, 123, 129
Television broadcasting. *See* Direct satellite broadcasting
Television, educational, 138–39
Tele-X satellite system, 133
Thermonuclear war, 37
Three Mile Island nuclear power plant, 87, 88, 90
Tools for Progress (ITDG catalogue), 19
Tracking and Data Relay Satellite, 132
Transportation, and intermediate technology, 23–25
Treaty of Friendship, Alliance, and Mutual Assistance (Sino-Soviet), 69, 75
Trudeau, Pierre, 89
Tsuruga nuclear power plant (Japan), 89
Twentieth Century Fund, 128

Uganda, satellite systems in, 126
Union of Concerned Scientists, 88–89
Union of Soviet Socialist Republics. *See* Soviet Union
United Arab Emirates, armaments in, 116
United Nations, 124, 140; UN Committee on the Peaceful Uses of Outer Space, 123; UN Conference on International Code of Conduct on the Transfer of Technology (Geneva, 1979), 56; UN Conference on Science and Technology for Development (Vienna, 1979), 58; UN Conference on Trade and Development (UNCTAD), x, 49–63 passim, 50, 53; UN Convention for the Recognition and Enforcement of Foreign Arbital Awards, 60; UN Economic Commission in Africa, 21; UNESCO, 25; UNIDO, 25; UN General Assembly, 53, 123
United Provinces. *See* Netherlands
United States: AID, 95; and arms sales, 118; and code of conduct for transfer of technology, 52–56; engineering profession in, 41–42; nuclear power plants in, 87, 89, 90; position of on Paris Convention of 1883, 58; Postal Service, 138; Small Business Administration, 107; satellite systems in, 122–23, 127, 129, 132–33, 135; space programs in, 122, 123, 126, 134–35; State Department Committee on International Investment, Technology, and Development, 55, 56
Uranium, production of, 91
United Steel Corporation, 104

Vance, Cyrus, 54, 88
Vietnam, South, 119
Village: global village, 121; village industries, 5, 9

Volcanoes, and nuclear power plant sites, 90
Volna satellite system, 133

Wallender, Harvey, 49
Wallerstein, Immanuel, 32
Weapons production, x, 37; in developing countries, 109–19 passim
Westar satellite system, 133
Western nations, and code of conduct for transfer of technology, 52–63
Western Union, 123, 133
Westinghouse Corporation, in Philippines, 83–96 passim

Wilgus, Walter, 93
Wilson, Stewart S., 10
Windpower, 27
World Administrative Radio Conference (WARC), 125, 139, 141
World Bank, 21–25, 27, 84, 95, 100
World Intellectual Property Organization (WIPO), 58
World War II, effects of on Korea, 99

Zaire, satellite systems in, 123, 126
Zambia, ITDG work in, 20, 21

Contributors

Warren E. Adams (Ph.D. in Economics, University of California at Berkeley, 1955) has been economic advisor to the Intermediate Technology Development Group, London, since 1976. Previously in the United States he was a faculty member at Earlham College, the University of Texas, and Swathmore College, and was a Carnegie Fellow at the Harvard Law School. Overseas he was a Ford Fellow and later taught at the Iraq College of Agriculture as well as serving as an economist with the USAID program in India. He has lectured, written, and done consultancies on rural development issues in many parts of the world.

Mathew J. Betz (Ph.D., Civil Engineering, Northwestern University, 1961) is currently Assistant Academic Vice President and Professor of Civil Engineering at Arizona State University. He has served on the faculty of the University of Khartoum, Sudan, and the University of Nairobi, Kenya. He has published widely in the area of transportation engineering and transportation planning regarding both domestic and international problems. He has served briefly in Nigeria, Ghana, and Egypt as well as the Sudan and Kenya. He has published in such journals as *Transportation Research Board Record* (National Academy of Science), *Traffic Quarterly*, *Transportation Research*, and the *Journal of Transport Economics and Policy*, as well as being the author of several monographs and research reports.

Daniel J. Bohlin, (Major, U.S. Air Force) is an instructor pilot and flight commander in the KC-135 Stratotanker with the 22nd Air Refueling Wing, March AFB, California. Following graduation from the U.S.A.F. Academy and flying duty for six years, he obtained an Olmsted Scholarship and studied political science for two years at the Institut d'Etudes Politiques, University of Grenoble, France. Returning to the United States, he gained his M.A. in Political Science from Arizona State University.

Werner J. Feld (Ph.D., Tulane University, 1962) is Professor of Political Science and Chair of the Department of Political Science at the University of New Orleans. Combining extensive experience in international business with a specialized study of international law and organization, he has written widely on issues of international political economy and the role of the United Nations Organizations in the elaboration of international regimes, including the Law of the Sea and technology transfer.

S. Ivy Lang (M.A. in Political Science, Arizona State University, 1982) is currently a graduate student in the Department of Political Science at A.S.U.

Pat McGowan (Ph.D. in Political Science, Northwestern University, 1970) is Professor and Chair of the Department of Political Science at Arizona State University. He has been a faculty member at Syracuse University and the University of Southern California. Before his graduate studies he was in the Peace Corps in East Africa (Kenya and Uganda) in the early 1960s. He has published widely on the comparative study of foreign policy, African international relations, and international political economy in such journals as *World Politics*, *The American Political Science Review*, *International Organization*, *International Studies Quarterly*, and *The Journal of Modern African Studies*.

Melba D. Solidum received an A.B. (cum laude) in Political Science from the University of the Philippines in 1976 and an M.A. in Political Science from Arizona State University in 1982. She is planning a career in the diplomatic service of her country.

Martin H. Sours (Ph.D. in Political Science, University of Washington, Seattle, 1971) is Professor of International Studies at the American Graduate School of International Management (Thunderbird). He has been a visiting professor in Korea (1973) and Japan (1977) and a visiting scholar in Taiwan (1982). He has published in the areas of U.S. foreign policy in Asia, Pacific regionalism, and Japanese management studies. His current professional interests include international political economy and the impact of service industries on the Pacific region. Before graduate studies he was a naval officer and currently holds the rank of Commander in the U.S. Naval Reserve.

Rolf T. Wigand, currently Professor of Public Affairs and Communication at Arizona State University, studied business administration at the Free University of Berlin and received his Ph.D. in Communication from Michigan State University. He has taught at Michigan State University and the Universidad Iberoamericana, Mexico City, and was a Research Fellow for the National Science Foundation. He has conducted grant research for the National Science Foundation, the National Institute of Mental Health, the International Social Science Council, the Defense Civil Preparedness Agency, and others. In the past he has consulted for Chase Manhattan Bank, AT & T, the U.S. Air Force, Honeywell LISD, IBM de México, Ford World Headquarters, and others. Presently he is editor of *Communications—International Journal of Communication Research*; he is past editor of *Systemsletter* and *Communicontents*. His writings have appeared in *Social Networks*, *Journal of Communication*, *Communications*, *Media-Perspektiven*, *Human Organization*, *Communication Yearbook*, and others. His current research interests are in the areas of the impact of information technology, social and communication networks, organization change, transborder data flow, and telecommunication policy.

Robert Youngblood (Ph.D. in Political Science, University of Michigan, 1972) is an Associate Professor of Political Science and a Research Associate of the Center for Asian Studies at Arizona State University. He has been engaged in research on the Philippines since 1963 and was the recipient of a Senior Fulbright-Hays Research Fellowship to the Philippines in 1979. He has published articles on the Philippines in journals such as *Asian Survey*, *Pacific Affairs*, *Comparative Political Studies*, *Australian Outlook*, the *Philippine Journal of Public Administration*, and the *Journal of Asian and African Studies*.

RAYMOND H. FOGLER LIBRARY

DATE DUE

BOOKS ARE SUBJECT TO
RECALL AFTER TWO WEEKS